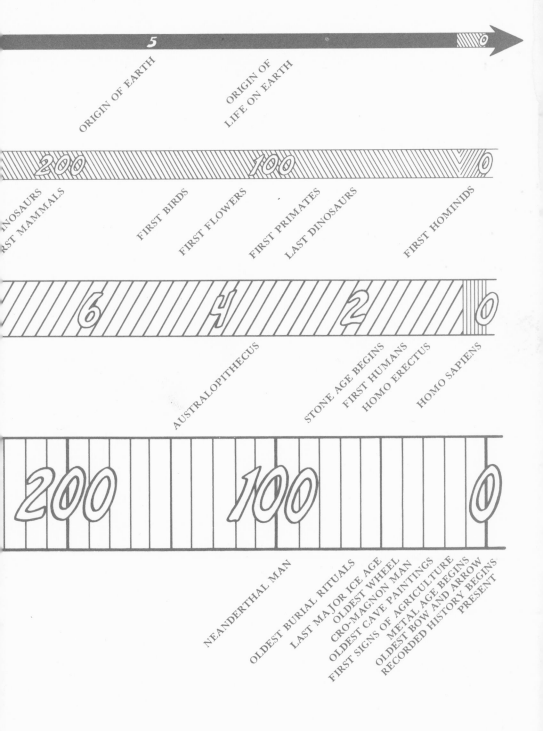

5

ORIGIN OF EARTH

ORIGIN OF
LIFE ON EARTH

200   100   0

DINOSAURS

FIRST MAMMALS   FIRST BIRDS   FIRST FLOWERS   FIRST PRIMATES   LAST DINOSAURS   FIRST HOMINIDS

6   4   2   0

AUSTRALOPITHECUS   STONE AGE BEGINS   FIRST HUMANS   HOMO ERECTUS   HOMO SAPIENS

200   100   0

NEANDERTHAL MAN   OLDEST BURIAL RITUALS   LAST MAJOR ICE AGE   OLDEST WHEEL   CRO-MAGNON MAN   OLDEST CAVE PAINTINGS   FIRST SIGNS OF AGRICULTURE   METAL AGE BEGINS   OLDEST BOW AND ARROW   RECORDED HISTORY BEGINS   PRESENT

COSMIC
DAWN

COSMIC DAWN

The Origins of Matter and Life

ERIC CHAISSON

Illustrated by Lola Judith Chaisson

An Atlantic Monthly Press Book

Little, Brown and Company — Boston–Toronto

LIBRARY OF CONGRESS CATALOGING IN PUBLICATION DATA

Chaisson, Eric.
    Cosmic dawn.

    "An Atlantic Monthly Press book."
    Bibliography: p.
    1. Cosmology. 2. Life — Origin.
QB981.C4   523.1   80-27828
ISBN 0-316-13590-9

ATLANTIC–LITTLE, BROWN BOOKS
ARE PUBLISHED BY
LITTLE, BROWN AND COMPANY
IN ASSOCIATION WITH
THE ATLANTIC MONTHLY PRESS

Published simultaneously in Canada
by Little, Brown & Company (Canada) Limited

PRINTED IN THE UNITED STATES OF AMERICA

To our parents

Preface

WHEN CONSCIOUSNESS DAWNED among the ancestors of our civilization, men and women perceived two things. They noted themselves, and they noted their environment. They wondered who they were and where they came from. They longed for an understanding of the starry points of light in the nighttime sky, of the surrounding plants and animals, of the air, rivers, and mountains. They contemplated their origin and their destiny.

Thousands of years ago, all these basic inquiries were treated as secondary, for the primary concern seemed to be well in hand: Earth was presumed to be the stable hub of the Universe. After all, the Sun, Moon, and stars all appear to revolve around our planet. It was natural to conclude, not knowing otherwise, that home and selves were special. This centrality led to a feeling of security or at least contentment — a belief that the origin, operation, and destiny of the Universe were governed by something more than natural, something supernatural.

Indeed, the ancients thought and thought. Their efforts produced such notable endeavors as myth, religion, and philosophy.

The idea of Earth's centrality and the reliance on supernatural beings were shattered only a few hundred years ago. During the Renaissance, humans began to inquire more critically about themselves and the Universe. They realized that thinking about nature was no longer sufficient. Looking at it was also necessary. Experiments became a central part of the process of inquiry. To be effective, hypotheses had to be experimentally tested, in order to refine them if experiment favored them or to reject them if it did not. The scientific method was born — probably the most powerful technique ever conceived for the advancement of factual information. Modern science had arrived.

Today, all physical and biological scientists throughout the world utilize the scientific method. They usually gather some data, then form a hypothesis, and finally test that hypothesis. This is a rational investigative approach used to formulate a description of all natural phenomena. Applied properly, the scientific method enables anyone to arrive at a conclusion free of the personal viewpoint of any one scientist. It's designed to yield a strictly objective consensus on the many aspects of our Universe.

People still query along the same lines as did the ancients. We ask the same fundamental questions: Who are we? Where did we come from? What is our origin and our destiny? But our attempts to answer them are now aided by the experimental tools of modern technology: astronomical telescopes to improve our vision of the macroscopic Universe of planets, stars, and even larger assemblages called galaxies; biological microscopes to aid our view of the microscopic world of cells, molecules, and even smaller entities called atoms; high-energy accelerators to probe the subatomic domain of nuclei, electrons, and even more elementary units called particles; man-made spacecraft to gather data unavailable from our vantage point on Earth; sophisticated computers to keep pace with the tremendously increased flow of new data, hypotheses, and experimental tests.

We live in an age of technology. And even though technology threatens to doom us, that same technology is now providing us with an unprecedentedly rich view of ourselves and our Universe.

Of all the scientific accomplishments since Renaissance times, one discovery stands out most boldly: Our planet is neither central nor special. Application of the scientific method has demonstrated that, as living creatures, we inhabit no unique place in the Universe at all. Research, especially within the past few decades, strongly suggests that we live on what appears to be an ordinary rock called the Earth, one planet circulating an average star called the Sun, one star in the suburbs of a much larger assemblage called the Milky Way Galaxy, one galaxy among countless billions of others distributed throughout the observable abyss called the Universe.

In this, the last quarter of the twentieth century, experimental science is now helping to unravel the details of the big picture. We are beginning to appreciate how all objects — from quarks to quasars, from galaxies to people — are interrelated. We are attempting to decipher the scenario of cosmic evolution: a grand synthesis of a long series of gradual alterations of matter, operating over almost incomprehensible space and incomprehensible time, that have given rise to our Galaxy, our Sun, our planet, and ourselves.

Modern science stipulates that steady change has been the hallmark for the development of all things. Researchers now have a reasonably good understanding, not only of how innumerable stars were born and have died to create the matter composing our world, but also of how life has come to exist as a natural consequence of the evolution of matter. There now seems to be a clear thread linking the evolution of simple atoms into galaxies and stars, the evolution of stars into heavy elements, the evolution of those elements into the molecular building blocks of life, of those molecules into life itself, of life into intelligence, and of intelligent life into cultured and technological civilization.

∞

To answer the fundamental questions Who are we? and Where did we come from? it's necessary to penetrate far back into the past, beyond the seventy years of an average human lifetime, beyond the start of modern science centuries ago, beyond the onset of language and civilization tens of thousands of years ago, beyond our ancestral man-apes who emerged from the forests several million years ago, even beyond the time when multicellular life began to flourish on our planet about a billion years ago — some ten megacenturies before now.

To appreciate cosmic evolution, we must broaden our horizons, expand our minds, and visualize what it was like long, long ago. We must go way back into the past. Go back, for instance, five billion years, when there was no life on planet Earth. Why? Because there was no Earth. There was no Sun, no Solar System. These objects were only forming out of a giant, swirling gas cloud at one edge of a vast galaxy of older stars that had already existed in one form or another for a long time before that.

Modern science now combines a wide variety of curricula — astronomy, physics, chemistry, biology, geology, anthropology, physiology, sociology, among others — in an interdisciplinary attempt to unravel the two most fundamental problems of all: the origin of matter and the origin of life. If we can understand the scenario of cosmic evolution, then perhaps we can determine precisely who we are, specifically how life originated on this planet, and, incredibly enough, how living organisms have evolved to the point of invading land, generating language, creating culture, devising science, exploring space, and even studying themselves.

∞

These writings concern all these things: space and time, matter and life. We explore our Universe, our planet, ourselves. We summarize where science stands today regarding answers to some of the time-honored philosophical questions — Who are we? Where did we come from? How do we, as living creatures, relate to the rest of the Universe? Where are we, as intelligent beings, likely to be headed in the future? In short, what are our origin and our destiny? What are the origin and destiny of the Earth, the Sun, the Universe?

Written for eclectic individuals having an interest in nature, this book explains valid contemporary science in a nontechnical manner. Accuracy has not been sacrificed, however, and a feeling for the frontiers of science has been included. Readers will recognize that answers to the above fundamental questions are not yet crystal clear. Even among technical peers, scientists are often unable to provide precise and detailed solutions for the broad and profound questions. Only within the past couple of decades have we gained the technological tools necessary to transfer these queries from the realm of philosophy to that of science. Researchers are now finding that the cutting edge of knowledge resembles a thinning haze rather than a real edge. The reason is that science is now moving at a fast clip, acquiring new knowledge at a phenomenal rate, and requiring novel interdisciplinary ventures to put it straight. Furthermore, much of it involves the human condition. As a fair assessment, we might say that a pencil sketch of the answers to the fundamental questions is now at hand, but that many details have yet to be furnished.

In a descriptive and illustrative way, then, we probe here the deepest nature of the Universe. These pages render the prevailing scientific view that the atoms in our bodies relate to the Universe in general. They elucidate the contemporary Universe view known as cosmic evolution — a truly big picture, a cosmogenesis, a dialectical materialism — whereby gradual changes in the composition and assembly of matter have given rise to galaxies, stars, planets, and life. We attempt to synthesize the essential ingredients of astrophysics and biochemistry, for these two subjects more than any others are making an enormous impact on our philosophical conception of ourselves and of our place in the Universe.

In short, this book presents the broadest view of the biggest picture. It analyzes, using the best science available, the most fundamental questions of all — neither the most relevant nor the most important questions, perhaps, for twentieth-century society, but the most fundamental ones. We develop an appreciation for our rich Universal heritage. We seek to unravel the nature and behavior of radiation, matter, and life on the grandest scale of all. We decipher the fabric of nature and, in the process, discover that technological humans now reside at the dawn of a whole new era.

∞

To make the scenario of cosmic evolution readable for a general audience, I've avoided referring in the text to the various authori-

ties. To quote each of the specialists would have detracted from the general concepts stressed throughout. The reader should recognize, though, that the knowledge described here was discovered by legions of researchers working across the entire spectrum of human understanding. The bibliography at the end of the book lists a sampling of many fine works I found useful while synthesizing my view of the big picture. It may be consulted for further reading.

My best coach in this undertaking has been my wife, Lola, who drew all the illustrations for this work. Her attempt to combine aspects of matter and life in the art form has been a source of considerable inspiration to me.

I am indebted to George Field, director of the Harvard-Smithsonian Center for Astrophysics, for inviting me to join him in teaching an interdisciplinary course on the subject of cosmic evolution to undergraduates of Harvard and Radcliffe colleges. The deep curiosity of these students to know more about our cosmic roots has helped to crystallize my thoughts on the grand synthesis sketched here.

Peter Davison, my editor at the Atlantic Monthly Press, contributed to the clarity of the manuscript and provided much assistance to a rookie author attempting to explore the intricacies of the publishing world.

Mark Stier and Ann Najarian made useful comments on an earlier version of the manuscript.

None of these acknowledgments, however, necessarily comprises an endorsement by any of the above people regarding the central themes I've chosen to stress in this Universal history.

ERIC J. CHAISSON
Summer, 1980
Winchester, Massachusetts

Contents

COSMIC DAWN

PROLOGUE

THE BIG PICTURE

EXPLORING THE WHOLE UNIVERSE requires large thoughts. There are no larger thoughts than cosmological ones. Cosmology is the study of the origin, evolution, and destiny of that aggregate of all matter and all energy known as the Universe. Here we strive to gain an appreciation for the bulk properties of the entire Universe: its current size, its shape, its structure.

Cosmic considerations must have a proper perspective. In considering the bulk properties of the whole Universe, the smaller contents such as planets and stars — even galaxies, to a certain extent — become irrelevant. Cosmologists regard planets as negligibly important, stars as only point sources of hydrogen consumption, and galaxies as mere details in the much broader context of all space.

Time also shrinks in significance when compared to eternity. An interval of a million years becomes a wink of an eye in the cosmic scheme of things. Even a billion years encompasses a rather short interval in the context of all time.

To appreciate cosmology, we must broaden our view to include all of space and all of time. If you've ever wanted to think big, now is the time!

At the very outset, take note: Thousands, millions, billions, and even trillions of things can be used easily in words. Not only are these truly enormous numbers, but the differences among them are also large. For example, one thousand seems easy enough to understand; at the rate of one number per second, we could count to a thousand in about fifteen minutes. However, to reach a million requires about two weeks, counting at the rate of one number per second, sixteen hours a day (allowing eight hours a day for sleep). And a count from one to a billion, at the same rate of one number per second for sixteen hours a day, would take an entire lifetime. A whole lifetime is required just to count to a billion! Yet we'll routinely consider here time intervals spanning millions and billions of not only seconds but also years. And we'll discuss objects housing millions of atoms, even billions of whole stars. Hence, we must become accustomed to gargantuan numbers of things, enormous intervals of space, and extremely long durations of time. Recognize in particular that a million is much larger than a thousand, and a billion much, much larger still.

∞

Viewing the Universe from our vantage point at Earth, we see a rich variety of objects. Among them are gassy nebulosities glowing with colorful light, explosive stars ejecting matter and energy, and powerful galaxies flickering in the depths of space. Through a telescope on a dark, moonless night, every object constitutes a superb example of astronomical architecture — a real jewel of the night. But astronomical objects are more than works of art, more than things of elegance. Planets, stars, nebulae, novae, galaxies, quasars, and all the rest are of vital significance if we are to realize our place in the big picture. Each is a repository of information concerning the material aspects of our Universe.

Light is only one type of radiation. Radio, infrared, ultraviolet, X-ray, and gamma-ray waves all comprise invisible radiation. But regardless of the type, radiation is energy. It is also information — a most primitive form of information. Yet, it's only by means of this one-way information flow that we can hope to fathom the depths of space.

Practitioners of astrophysics acquire information about cosmic objects by interpreting the radiation they emit. We say "astrophysics" because that word defines more than any other the basis

4

on which the interpretation is made. The emphasis is on physics; *astro* is a mere prefix. The space scientist of today who doesn't have a firm grounding in physics is no space scientist at all. Gone are the days for astronomers to make fundamental discoveries by peering through telescopes and marveling at the sights. The modern astrophysicist wants to know more than just where objects are, or what their brightness and color may be. We want to perceive *what* lurks beyond the range of eyesight. We seek to understand *how* the myriad objects got there, *how* they operate, and especially *how* matter and radiation interact. In sum, astrophysicists aspire to comprehend the origin, evolution, and destiny of all that lies beyond planet Earth.

There's an essential difference between the majority of scientists, who study terrestrial matter in a laboratory on Earth, and astrophysicists, who investigate extraterrestrial matter that is far away from our home planet.

On Earth, scientists can control their experiments as an aid to discovering the properties of all types of terrestrial matter. This control can be exercised either by manipulating the matter under scrutiny or by tinkering with experimental techniques used to inspect it. Consider, for example, an attempt to unravel the properties of some new rocky ore. Laboratory scientists could utilize a variety of rock samples, each having a different shape and size. They could alter the orientation of the rocks within the laboratory apparatus. They could vigorously heat the ore or cryogenically cool it, and even subject it to varying amounts of electricity and magnetism. All the while, researchers can learn a great deal about the mineral ore by deciphering its responses to these environmental changes. In short, the environment in which a terrestrial experiment operates can be altered or manipulated in order to enhance the study of any piece of local matter.

Matter far beyond our planet, however, cannot be massaged — not even with the very best tools of modern civilization. Distant extraterrestrial environments cannot be controlled or manipulated. Astrophysicists are confined to working with intangible radiation emitted by extraterrestrial matter — radiation occasionally intercepted by the mind's eye or detected by instruments on Earth, signals momentarily captured while they are traveling from distant objects to faded oblivion someplace in the dim recesses of the Universe.

Of course, technological advances have recently provided a few exceptions to the above statements, enabling space scientists to perform guided experiments on a few specimens of nearby extraterrestrial material: interplanetary meteorites discovered buried in Earth's crust and especially in its icy polar regions, lunar rocks retrieved from our dead neighbor via the American and Russian space programs, and Martian soil examined by a couple of robot spacecraft now parked on the plains of that rusty planet. Yet it's likely to be many centuries before our descendants have the capability to conduct in situ exploration of matter exterior to the system of planets now familiar to everyone. For now, the bulk of Universal matter must be inventoried and analyzed by extracting information veiled within its naturally emitted radiation captured by our telescopes on or near Earth.

Details aside, radiation is the only means whereby we know of the existence of celestial objects.

There's a further restriction when contemplating distant extraterrestrial matter. Not only are we prohibited from studying celestial objects at their locations in space, but we are also denied the chance to examine them now in time. Why? Because radiation does not travel infinitely fast; it travels at a finite speed — the velocity of light. Consequently, it takes time — often lots of time — for light or any type of radiation to travel the unimaginably vast expanses of space separating objects in the Universe.

Consider as an example the nearest star (save the Sun), called Proxima Centauri, a member of the triple star system Alpha Centauri. Despite the adjective *nearest*, this star is still four light-years away — a terribly long range once we realize that a light-year is the *distance* traveled by light in a full year at the fastest velocity known.

A light-year, then, is a distance. It's equivalent to about ten trillion kilometers, or six trillion miles. (That breaks down to some thirty billion kilometers traveled each day by light or any other type of radiation. Fast, there is no doubt.) Yet the Alpha Centauri system is not just a single light-year away from us; it is several light-years distant. Accordingly, radiation takes several years to travel from this star system to Earth. Since nothing can surpass the velocity of light, Centauri's radiation simply could not get here any quicker. Expressed another way, the light we see while looking at Proxima Centauri today left that star several years ago. It's been cruising through the near void of outer space ever since.

Radiation from distant objects, therefore, harbors clues to the past. The farther an object is from Earth, the longer its light takes to reach us. Radiation emitted by the nearest galaxies, millions of light-years distant, left those objects before *Homo sapiens* emerged as part of the animal family on planet Earth. Radiation observed from truly distant galaxies actually left those objects well before Earth even formed. Indeed, radiation now reaching Earth from the most remote cosmic objects was launched in earlier epochs of the Universe when there were no planets, no stars, no Milky Way Galaxy.

By studying radiation, astrophysicists can learn what the conditions were like long ago when distant objects emitted their light. Deciphering hints within that radiation, we can not only imagine the general conditions in the Universe before the creation of the Sun and Earth, but we can also specify values for the two most important factors — temperature and density — characterizing the Universe in those ancient times.

Our perspective of the Universe is delayed. We see the Universe as it was, not as it is.

Astrophysicists, then, are the ultimate historians. Looking out from Earth, we see a history of the Universe. Telescopes are our time machines — tools enabling us to probe earlier phases of the Universe, including aspects of our origins. Much like archeologists who dig through ancient rubble in search of hints about the origin and evolution of cultures, astrophysicists interpret radiation, seeking clues about the origin and evolution of matter itself.

So never forget: *Looking out into space is equivalent to looking back into time*. By examining deep space and capturing radiation from the most distant objects, researchers hope to procure a portrait of what the Universe was like long ago, including near the time of its origin. This is the task before us.

∞

Cosmic activity permeates our Universe. So does quiescence. Perspective determines which dominates. Examined casually, astronomical objects usually display stability. Higher resolution, however, often reveals some violence. Generally, the larger the perspective, the more stable things seem. For example, that our Earth is ruptured by quakes and volcanoes is obvious to those of us who live on it and witness its daily activity up close; but our planet appears tranquil when viewed from afar in those striking lunar earthrise photos taken by the Apollo astronauts. Similarly,

8

telescopic studies of our Sun show it to be peppered with bright flares, dark spots, and surface explosions, as are, presumably, all stars; but to the naked eye, the Sun and most stars assume a rather peaceful, steady pose.

We might expect that, while there will surely be pockets of violence tucked here and there throughout the fabric of the Universe, the largest possible perspective would give rise to a model of perfect quiescence. Not so, however. In bulk, the Universe is not calm and stable. Surprisingly, the whole Universe itself displays considerable activity.

Once we realize that the Universe harbors a certain verve, we might then expect the largest material coagulations — among them the galaxies — to have random, disordered motions, some hurtling this way, and some that. Chaotic motions of fireflies trapped in a large jar come to mind, or even turbulent motions of peas in a pot of boiling soup. For the Universe, however, these are not good analogies. Galaxies are not moving chaotically. The Universe is indeed active, but in an awesomely ordered fashion.

For fifty years now, scientists have been aware that galaxies have some definite organized movement — a Universal traffic pattern of sorts. Peculiarly enough, virtually all galaxies seem to be steadily receding, propelled away from us as though we had a kind of cosmic plague. Not only that, but they are also receding according to a grand overall design. Each galaxy drifts away at a velocity proportional to its distance from Earth. This is a fact of great significance: the greater the distance of an object from us, the faster that object recedes. There is a linear relationship — a perfect correlation — between velocity and distance.

The entire pattern of distant objects receding more rapidly than nearby ones implies, if we think about it for a moment, that an explosion must have occurred at some time in the past. The more distant an object is from us, the more forcefully it — or whatever formed it — must have been initially expelled. This is precisely the flight pattern of shrapnel fragments when a conventional bomb explodes. There is no need to postulate, indeed there is absolutely no evidence for, a repulsive force pushing the galaxies apart. Instead, the galaxies are simply the debris of a primeval explosion, a cosmic bomb whose die was cast long ago.

The recessional motion of all the galaxies demonstrates that the entire Universe is quite active. On the largest scale of all, the Universe itself is in motion. It's not at all a pillar of stability. Instead, the Universe is changing with time. It's expanding, and expanding in a directed fashion — in short, evolving. This does not mean that either the Solar System or the individual galaxies are physically ballooning in size. Planets, stars, and galaxies are all gravitationally bound, dynamically intact entities. Only the motions of the largest coagulations of matter share in this Universal expansion. Accordingly, distances separating galaxies and clusters of galaxies grow larger with time.

Astrophysicists, philosophers, theologians, as well as people from all segments of society would like to know if the Universe will continue to expand in this way forever, or if it will eventually stop. We'd also welcome more information about the nature of the explosion that gave rise to the galaxy motions. If the Universe eternally expands, there will be unimaginable time available for the continued evolution of matter and life. Alternatively, should the Universe embody enough matter, the combined pull of gravity could bring the expansion to a halt, and even reverse it into contraction.

This latter possibility brings several questions to mind: How much future time will elapse before the Universe ceases expanding? If it does start to contract, what will happen upon eventual collapse of the entire Universe? Will the Universe simply end as a small, dense point much like that from which it began? Or will it bounce and begin expanding anew? Perhaps the Universe has rebounded in this way before. Maybe we inhabit a cyclically expanding and contracting Universe — one having a continuous array of cycles, though never a true beginning or end.

These, then, are the basic large-scale fates of the Universe in bulk: It can expand forever. It can expand and then contract to a

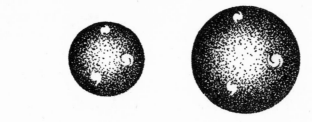

virtual point and end. Or it can cyclically expand and contract indefinitely. Each model represents a hypothesis — a theory based on available data. But unless we can take that third step in the scientific method and experimentally test them, we cannot know which version is correct, if any.

As tricky a task as this may seem, these various models are currently being subjected to observational tests by contemporary astrophysicists. Their experiments, and the theories underlying their experiments, seek direct answers to many of the above questions. Even a superficial appreciation of the current status of their solutions, though, requires a deep understanding of the nature of space and time. To gain this understanding, we need a particular tool. And that tool is relativity theory.

∞

Some people become hot, bothered, and tense upon hearing the word *relativity*. But, conceptually, relativity theory is relatively simple. Its foundations are rather straightforward, provided we are willing to forgo common sense and human intuition.

Relativity is simple in its symmetry, its beauty, its elegant ways of describing grandiose aspects of the Universe. Sure, it employs higher mathematics — advanced calculus and beyond — to quan-

tify its application to the real Universe, yet everyone should strive to gain at least a nonmathematical feeling for the underlying concepts of relativity theory. Such an understanding is the basis for an appreciation, albeit only a qualitative one, of some of the weird astrophysical effects to be encountered while studying the origin of the Universe, exploring the vicinity of black holes, and modeling the entire Universe.

Relativity theory has two principal tenets, both enunciated in 1905 by the German-American physicist Albert Einstein; together they lead to the famous $E = mc^2$ equation, where E, m, and c are symbols representing energy, mass, and light velocity, respectively. The first tenet maintains that the laws of physics are the same everywhere and for all observers. Regardless of where a person is, or how fast a person may be moving, the basic physical laws are invariant.

The second tenet of relativity is that there is a fourth dimension — time — which is in every way equivalent to the usual three spatial dimensions. There are three dimensions of space; an object's position can be generally described as either left or right, either up or down, and either in or out. Three dimensions are sufficient to describe *where* any object is in *space*. A fourth dimension of *time* is necessary to describe *when* — either past or future — an object exists in that space. By coupling time together with the three dimensions of space, Einstein was able to reconcile previous inconsistencies in Isaac Newton's Renaissance view of our world by hypothesizing that the velocity of light is an absolute constant number at all times and to all observers, regardless of when, where, or how radiation is measured. Space and time are in fact so thoroughly intertwined within Einstein's view of the Universe that he urged us to regard these two quantities not as space *and* time, but as one — *spacetime*.

A number of important consequences of relativity theory can be qualitatively explained only by analogy. Here is one of them: Suppose we are in an elevator having no windows. When it rises, we feel the floor pulling, especially on our feet. It's easy to attribute this pulling sensation to the upward acceleration of the elevator. Now, imagine that such a windowless elevator exists in outer space far from Earth. Normally, we would experience the weightlessness made familiar by watching the astronauts floating around where there are no strong gravitational forces. But if we *did* experience a sensation of pulling on our bodies, we could conceive one of two possible explanations. We could argue that the elevator is accelerating upward in the absence of gravity, thus pinning us to the floor. Or we could maintain that the elevator is at rest in the presence of gravity, which is pulling us from below. There is no way to tell which of these explanations is correct without performing experiments on the outside world — that is, without looking at objects outside the hypothetical elevator. If we could look out, we would have no trouble establishing whether the elevator is really at rest or really accelerating. Relative to the Universe outside the elevator, it's easy to determine the real status of that elevator.

The important point is that the effect of gravity on an object and the effect of acceleration on an object are indistinguishable. Scientists call this the Principle of Equivalence: Gravity and the acceleration of objects through spacetime can be viewed as conceptually and mathematically equivalent. Consequently, Einstein postulated as unnecessary the Newtonian view of gravity as a force that pulls. Not only is that view unnecessary, but Newton's theory is today known to be less accurate than Einstein's.

Let's briefly examine how the notion of accelerated objects can replace the commonsense idea of gravity. General relativity allows us to inquire how it is that matter, which, of course, normally gives rise to the Newtonian view of gravity, alters the nature of spacetime. Bypassing the details, the upshot is that matter curves or warps spacetime. In other words, matter itself effectively shapes spacetime.

Flat Euclidean geometry, the type learned in high school, holds valid when the extent of curvature is zero. Even when the extent of curvature is slight, Euclidean geometry of flat space is approximately correct. At any one location on Earth's surface, for instance, an architect can design a building, or a contractor can construct

one, using the normal procedures laid down twenty-five centuries
ago by the Greek mathematician Euclid. Even though flat-space
geometry is used every day by many workers, it's not absolutely
correct. It can't be correct. The Earth, after all, is not flat. It's
curved. Flat Euclidean geometry works satisfactorily at any small
locality. But that's because it's impossible to perceive our planet's
curvature from any single surface locality. Once the curvature of
Earth becomes a factor, as in the case of intercontinental aircraft
navigation for example, then a more sophisticated geometry must
be used — a curved-space geometry.

And so it is at selected locations in the Universe; in the absence
of matter, the curvature of spacetime is zero, and an object in that
kind of flat space moves uniformly in a straight line. Newtonian
dynamics and Euclidean geometry are valid, for all practical purposes, when spacetime is not appreciably curved. Flat space is not
a hypothetical situation, for, beyond the reaches of galaxies, very
little matter presumably exists, and thus hardly any warping of
space is expected.

On the other hand, the geometry of spacetime is strongly warped
near massive objects. It's not the object or the surface of the object
that's warped — just the near-void of spacetime in which the object is embedded. The larger the amount of matter at any given
location, the larger the extent of curvature or warp of spacetime
at that location. At greater distances from the location of a massive
object, the extent of warp lessens. As with gravity, the extent of
spacetime curvature depends upon both the amount of matter and
the distance from that matter. But, since this view of warped spacetime is more accurate than the conventional view of gravity, to be
correct precisely in all possible situations, the Universal view of
Newton must be replaced by that of Einstein.

Yet, surely you ask: How can a curve replace a force? The answer
is that the topography of spacetime influences celestial travelers in
their choice of routes much as Newton imagined gravity to hold
an object in its path. Just as a pinball cannot traverse a straight
path once shot along the side of a bowl, so the shape of space
causes objects to be accelerated. They don't move uniformly along

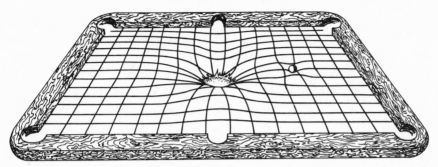

straight lines; instead, they move along curved paths. For example, Earth is accelerated in its orbit about the Sun — not because of gravity, as Newton would explain, but because of the curvature of spacetime, as Einstein would prefer. As inhabitants of Earth, we don't feel our planet accelerating, but it is; any object that changes its direction of motion is actually accelerating.

To appreciate further the Einsteinian view of spacetime, consider an analogy. Imagine a pool table with a top made of a thin rubber sheet, rather than the usual hard felt. Such a rubber sheet would become distorted if a large weight were placed on it. A heavy rock, for instance, would cause the sheet to sag, to warp. The otherwise flat rubber sheet would become curved, especially near the rock. The heavier the rock, the larger the curvature. Trying to play billiards, we would quickly find that balls passing near the rock were deflected by the curvature of the table top.

In much the same way, both radiation and material objects are deflected by the curvature of spacetime near a massive object. The Earth's orbital path is deflected from moving in a straight line by the slight spacetime curvature created by our Sun. The extent of the deflection is enough to cause our planet to circle the Sun. Similarly, the Moon or a baseball responds to the spacetime curvature created by the Earth, and accordingly moves along a curved path. The deflection of the Moon's path is not too large, causing our neighbor to orbit Earth endlessly. The deflection of a small baseball's path is much larger, causing it to return to Earth's surface.

The notion of gravity, then, is nothing more than the natural

15

behavior of objects moving within the geometrical framework of a warped spacetime. As a result, it's possible to use a knowledge of spacetime in order to predict the motions of objects through space and time. Greater usefulness can be gained by turning the problem around: By studying the accelerated motion of any object, we can learn something about the geometry of spacetime near that object.

And so it is with the whole Universe. When we seek the size, shape, and structure of the entire Universe — the biggest picture of all — it's necessary to consider the accumulated effects of spacetime curvature near every massive object throughout the cosmos. By studying the accelerated motion of all the matter within the Universe, we can hope to learn something about the curvature of the whole Universe.

Infusing relativity's basic tenets throughout a full-blown development of field theory makes possible the mapping of specific ways in which matter warps spacetime. This is the area where relativity theory becomes notoriously complex; here, theorists scamper away, leaving us in the qualitative dust. What we glean from their ponderous calculations can only be subjective. The results in a nutshell are the so-called Einstein field equations. A dozen or so equations that must be solved simultaneously, they determine how the Universe is structured — that is, how spacetime is curved by the matter present. Although on the one hand these equations are intractable, on the other hand they contain remarkable symmetry; much like a work of art, they are often accorded a sense of wonder, a certain awe. The complexity arises largely because, along with these field equations determining the geometry of the Universe, the relativist must additionally solve several geodesic equations to determine how it is that any given object behaves in and among all the other matter in the Universe.

To illustrate further the curvature of spacetime, ponder the following example. Suppose there are two planets, each inhabited by a civilization capable of launching rockets. Earth can be one, since we now have the required technology. For the other planet, imagine Mars to have an advanced technological civilization. Suppose, furthermore, that the inhabitants of both planets possess identical rockets. For the sake of discussion, let's assume that these rockets can achieve only a fixed amount of thrust at launch, after which they glide freely through space. When the rockets are launched from both planets, the shapes of their paths differ. In the Newtonian view of space, the rocket path is determined by the gravitational interaction between the rocket and each planet. In the Einsteinian view of spacetime, the rocket path is determined by the response of the rocket to the spacetime warp created by each planet.

Consider first a possible path of the rocket launched from the more massive planet, Earth. The initial kick is chosen in this case to be large enough to place the rocket into an elliptical orbit. Like gravity, whose strength decreases with increasing distance from a massive object, the curvature of spacetime is also greater close to the massive planet. The rocket accordingly speeds up (or accelerates) when close by, and slows down (or decelerates) when far away. General relativity thus agrees with the laws of planetary motion empirically discovered a few centuries ago by the German astronomer Johannes Kepler. Relativity maintains that the rocket moves faster (accelerates) when close to a massive object because the rocket is actually responding to the greater degree of spacetime curvature there.

The ellipse, a closed geometric path, is only one possible type of course. It's a trajectory of minimum energy, so labeled because the rocket having such an orbit does not possess enough energy to escape the planet's influence.

Rockets can have other paths as well. Consider the trajectory taken by an identical rocket upon launch from the less massive planet Mars. The same thrust used to launch the Earth rocket into elliptical orbit is great enough to propel the rocket entirely away from Mars. Less energy is used in the launch from Mars than in the one from Earth, and thus more energy can be imparted to the motion of the rocket. The rocket can literally escape the influence of Mars because, as a Newtonian classicist would say, Mars has less gravitational pull than Earth. Alternatively, Einsteinian relativists maintain that such a rocket escapes Mars because the less massive Mars warps spacetime considerably less than does Earth. The two views — Newtonian and Einsteinian — predict virtually identical paths for the rocket as it recedes toward regions of spacetime progressively less curved by Mars.

The resultant path away from Mars is called a hyperbolic trajectory. This is the type of flight path taken by our spacecraft that have been probing other planets in recent years. Ignoring all other astronomical objects that also curve spacetime, we can state that any object moving along such a hyperbolic path has enough energy to approach infinity. In general, an object traveling along a hyperbolic path possesses more energy than one on an elliptical trek. This is true either because the initial kick necessary to achieve a hyperbolic trajectory is large, or because the mass of the parent object from which the launch is made is small. In the particular example considered here, the rockets are identical; so the increased energy of the hyperbolic case results from the comparatively small mass of Mars.

Of course, even while receding far from its parent planet, a rocket will still be affected by the pull of gravity, or the warp of spacetime, created by the mass of that planet. Although substantial only in the immediate vicinity of the planet itself, Mars' influence over the rocket never diminishes to zero. Mathematical analyses show that, in the idealized absence of all other astronomical objects, such a hyperbolically launched rocket is predicted to reach infinity. But since nothing can ever really reach infinity, this is tantamount to the statement that the rocket will continue to recede indefinitely.

The above two cases comprise convenient ways in which trajectories of any object can be described in terms of their energy content and of response to the curvature of spacetime. They'll be useful analogies when we consider the essentials of cosmology, for then the "object" will be the whole Universe itself. We'll return to these ideas in a moment.

∞

Einstein, as the originator of relativity, was in a better position than anyone else to use his equations to determine the structure and nature of the entire Universe. His field equations predicted in 1917 that the curvature of the whole Universe must indeed be large. The flat geometry of Euclid just won't do when examining the bulk properties of the entire Universe. Unfortunately, Einstein's solution can be cast only in terms of unimaginable four-dimensional spacetime. It's quite imaginable mathematically, but it's totally unimaginable conceptually.

To visualize the essence of the solution, we can use an analogy. No one has ever built a viewable model of anything in four dimensions, so in this analogy, we suppress one of the four dimensions. For example, for the sake of argument, we can consolidate the three dimensions of space into only two dimensions of space. Then, with time as the remaining dimension, we can appreciate a three-dimensional analogue of Einstein's Universe.

The analogue is a sphere. All of space is considered to be distributed *on the surface* of this sphere. In other words, all three dimensions of space have been consolidated into two dimensions, and these two dimensions exist on the surface of the sphere. The

remaining dimension — time — is represented by the radius, or depth, of the sphere.

It's important to emphasize once again that, in our analogy, the Universe and its contents should *not* be considered to be distributed within the sphere. Rather, they are distributed *just on its surface*. All three dimensions of space are warped — in this special case, into a perfect sphere — simply because of the net influence of all the matter within every astronomical object. Thus, all the galaxies, all the stars, all the planets, and even all the radiation exist only on the surface of the sphere of this modeled Universe.

Now, since the radius of this model sphere represents time, we are forced to conclude that this spherical analogue grows with time. After all, the galaxies are observed to be receding. As time marches on, the radius of the sphere increases and so does its surface area. In this way, this three-dimensional analogue agrees with the observational fact that the Universe is indeed expanding.

* * *

Actually, Einstein didn't know in 1917 that the Universe is waxing. Astronomers did not establish the expansion of the Universe until the 1920s. Einstein's own field equations had predicted an expansion of the Universe, but he didn't believe it. He was probably fooled by the still-lingering Aristotelian philosophy that few things change. So he tinkered with the field equations, introduced a fudge factor that just offset the predicted expansion, and forced the model Universe to remain static. Einstein was wrong in doing this, of course. He later declared that it was the biggest blunder of his scientific career. But in the process, he and other relativists chanced upon many important features of curved spacetime. One of the most important findings is known as the Cosmological Principle — the notion that all observers perceive the Universe in roughly the same way regardless of their actual locations.

To grasp the Cosmological Principle, consider a sphere again. It can be any sphere, so let it be Earth. Imagine we are at some desolate location on Earth's surface, perhaps in the midst of the Pacific Ocean. To make the analogy valid, we must confine ourselves to two dimensions of space; we can look east or west, north or south, but we cannot look up or down. This is the life of a fictional "flatlander" — a person who can visualize only two dimensions of space. Perceiving our surroundings, we note a very definite horizon everywhere. The surface *appears* flat and pretty much the same in all directions. Accordingly, we might get the impression of being at the center of something. But we're not really at the center of Earth's surface at all. There is no center on the surface of a sphere. Such is the Cosmological Principle: There is no preferred or central location on the surface of a sphere.

Likewise, regardless of our position in the real four-dimensional Universe, we observe roughly the same spread of galaxies as would be noted by any other observer from any other vantage point in the Universe. Despite the observation of galaxies all around us, it doesn't necessarily mean that we reside at the center of the Universe. In fact, if our spherical analogy is valid, there is no center of the Universe. Nor is there any edge or boundary. The case of a flatlander coasting on the surface of a three-dimensional sphere is completely analogous to that of a space traveler voyaging through the real four-dimensional Universe. Neither one of them ever reaches a boundary or an edge. A flatlander will, by roaming the sphere far enough in a single direction, eventually return to the starting point. In much the same way, if four-dimensional space-time is structured according to this spherical analogy — and it may be — an astronaut could be launched in one direction, only to return at some future date from the opposite direction.

Today, we recognize that the Universe is not at all static. The recessional motions of the galaxies make this fact indisputable. Led by the 1920s efforts of the Russian mathematician Alexander Friedmann and the Belgian priest Georges Lemaître, modern relativists seek more realistic models of the Universe, especially ones that take account of the observed rate of Universal expansion. In this way, observations of galaxy recession become a restraint on any plausible model of the Universe, helping refine our late twentieth century view of the big picture.

One thing worth noting is that no place on the surface of an expanding sphere, like that of any static sphere, has a central location. The Cosmological Principle is valid even though the Universe is expanding. To see this, imagine a sphere again, though now one that can swell. For example, visualize the entire Earth to be expanding, causing the surface area of our planet to increase as time progresses. Standing on such a hypothetically expanding "Earth," we would see familiar objects receding; all surface objects on all sides, whether they were trees, houses, or mountains, would appear to recede. Now, more than ever, we may want to conclude that our position is special — that we exist at the center of some explosion. But we don't. Our position is no more special than anyone else's on the sphere's surface. In fact, everyone everywhere on the expanding surface would observe their surroundings to be receding. Who is correct, then? Everyone is correct. Recessional motions are observed from *any and all* positions on the surface of an expanding sphere.

Another common way of visualizing the same concept is to paint dots on the surface of a balloon. The dots are meant to represent the galaxies, and the balloon the Universe. As the balloon inflates, all dots recede from one another. Regardless of the galaxy we happen to inhabit, we would note that all other galaxies were receding. The galaxies appear to recede for any observer in the Universe. There's nothing special or peculiar about the fact that all the galaxies are receding from us. This is the case for all observers everywhere. Such is the Cosmological Principle: No observer anywhere in the Universe has a privileged position.

And so it is in the real four-dimensional Universe. Despite the fact that all galaxies recede from planet Earth, this is not a peculiarity of our vantage point. All observers anywhere in the Universe would see essentially the same sort of galaxy recession. Neither we nor anyone else is at the center of the expanding Universe. There is no real center in *space*. There is no position that we can ever hope to identify as the location from which the Universal expansion began.

There is nonetheless a center in *time*. This is the origin of time, and it corresponds in our three-dimensional spherical analogy of four-dimensional spacetime to the sphere's radius of zero. In other words, at the beginning of the Universe, the three-dimensional sphere was a point. It had zero radius. This was the origin of time. It's proper to think of it as the edge of time. But there is no edge in space.

It's reasonable to ask the obvious question: When did the sphere have zero radius — a mere point? In other words, how long ago were all the contents of the Universe concentrated into a single speck? Fundamentally put, when did time begin?

To appreciate answers to these questions, imagine that time can be reversed. We can mentally reverse the expansion of the Universe by contracting it at the same rate as we currently observe it expanding. The galaxies would come together, eventually they would touch, and finally they would mix. If we can estimate how long it would take for the whole Universe to shrink to a single point, we'll then have a measure of the age of the Universe.

The answer is approximately fifteen billion years. Thus, the singular, compact region of space often associated with the origin of the Universe must have existed about fifteen billion years ago. Another way of looking at it is that fifteen billion years have passed since the exploded debris of Universal matter reached the locations at which they are now observed.

The range of possible error in this number is considerable, for the true rate of galaxy recession is hard to measure with any degree of accuracy. Some researchers argue that the Universe could be as young as ten billion years, while others maintain it is nearly as old as twenty billion years. An error of several billion years may seem large, but the difference between these two extremes is only a factor of two, really pretty good for an order-of-magnitude subject like astrophysics. As a compromise, we adopt fifteen billion years as the approximate age of the Universe — a remarkable finding in and of itself.

∞

At the origin of time, the Universe began to expand. Like air blowing up a balloon, time pushes the Universe out into the future — the galaxies recede and the Universe expands. Actually, it expands at a rate inversely proportional to the density of matter contained within it. After all, each clump of matter in the Universe gravitationally pulls on all other clumps of matter. Since this gravitational force is always attractive, it tends to counteract the expansion. So a large amount of matter causes a substantial gravitational pull, and eventually a slowing of the Universal expansion. (Notice that we've returned to the notion of gravity; though warped space is a more valid concept, we'll use the familiar gravity whenever it makes the argument easier to understand.)

The phenomenon of Universal expansion is not unlike what happened with the rockets considered earlier. Each rocket moved away from its parent planet at a rate inversely proportional to the planet's mass. Mars, for example, pulled strongly on the launched rocket, but was unable to slow the rocket's escape; the more massive Earth exerted an even stronger pull on the rocket, and was able to halt its escape. The analogy between the orbital dynamics of a rocket and the cosmic dynamics of the Universe is quite a good one. Just as for rockets, there are essentially two possible models of a dynamic, changing Universe.

The first model Universe corresponds to one that evolved from a powerful initial "bang" — an explosion of some sort that occurred at the origin of time. The Universe then expanded from what must have been an exceedingly dense primeval clump of matter. As time progressed, space diluted matter throughout the Universe, causing the average density to decline. In this first model, there is insufficient matter to counteract the expansion. Accordingly, the Universe simply expands forever, with the density of matter thinning out eventually to nearly zero. This type of possible Universe will theoretically arrive at infinity with some finite (nonzero) velocity. It's specifically analogous to the rocket moving away from Mars; this type of Universe has insufficient mass to halt the matter's outward motion. Some researchers refer to this case as the hyperbolic model of the Universe.

A hyperbolic model is said to imply an "open Universe." It's open in the sense that the initial bang was large enough and the contained matter thin enough to ensure that this type of Universe will never stop expanding. Despite the fact that matter everywhere pulls on all other parts of the Universe, this type of Universe will never collapse back on itself. There's simply not enough matter.

Of course, the Universe can never really become infinitely large. An infinite amount of time is required to reach infinity. This is just the mathematician's way of stating that the hyperbolic, or open, Universe will continue to expand forever. Properly stated, the Universe approaches infinity.

Should this model be correct, the galaxies will recede forevermore. With time, for an observer on Earth, they will literally fade toward invisibility, their radiation becoming weaker with increasing distance. Eventually, even the closest galaxies will be so far away that they'll hardly be visible. Someday, they too will become unobservable; they will be too far away, their radiation too faint. The Milky Way Galaxy will then be the only matter within our observable Universe. All else, even through the most powerful telescopes, will be dark and quiet. And frighteningly enough, even beyond that in time, the Milky Way too will someday peter out as the hydrogen in all its stars becomes spent. This type of Universe, and all its contents, eventually experiences a "cold death." All the radiation, matter, and life in such a Universe are destined to freeze.

There is another plausible model for the Universe. As was the case with the open Universe, this model expands with time from a superdense original point. But unlike the open Universe, this model contains enough matter to halt the Universal expansion before it reaches infinity. That is to say, once the bang has initially pushed out the Universe, the galaxies recede ever more slowly, eventually skidding to a stop sometime in the future. Astronomers everywhere — on any planet within any galaxy — will then announce that the galaxy recession has subsided. The Cosmological Principle guarantees that this new view will prevail everywhere. The bulk motion of the Universe, and of the galaxies within, will be stilled — at least momentarily.

The expansion may well stop, but the inward pull of gravity never will. Gravitational attraction is relentless. Accordingly, this type of Universe will necessarily contract. It cannot stay motionless. Nothing fails to change. The contraction of this type of Universe is a mirror image of its expansion. Not an instantaneous collapse, it's rather a steady movement toward an ultimate end, requiring just as much time to fall back as it took to rise.

This type of model is in many ways analogous to the rocket trajectory for which, in our earlier example, the gravitational pull was great enough to cause the rocket path to become elliptical. Since it has a similar geometrical pattern, a Universe model containing enough matter to reverse the expansion is often called an elliptical

Universe. It's also sometimes referred to as a "closed Universe" — closed because it represents a finite Universe of finite size. It has a beginning and it has an end.

The variation of density in a closed Universe is interesting. From what must be an enormously large initial value, it thins to a rather small value by the time the Universe begins to contract, then returns again to an enormously large value when, at some distant epoch of the future, all the matter collapses onto itself.

The expansion-contraction scenario of a closed Universe has many fascinating (and dire) implications. Life, in particular, which has evolved steadily from simplicity to complexity during the expansion, will begin breaking down toward simplicity again while inevitably heading toward its demise during the contraction. Why? Because toward the end of the contraction phase, the galaxies will collide as the total amount of space in which they exist diminishes. And just as compressing a gas or rubbing our hands causes heating via friction, collisions among galaxies will generate heat as well. The entire Universe will grow progressively denser and hotter as the contraction approaches the end. Near total collapse, the temperature of the entire Universe will have become greater than that of a typical star. Everything everywhere will have become bright — so bright that stars themselves will cease to shine for want of contrasting darkness. This type of Universe, then, will shrink toward the superdense, superhot state of matter similar, if not identical, to the one from which it originated. In contrast to the open Universe that terminates as a frozen cinder, this closed Universe will experience a "heat death." The contents of this Universe are destined to fry.

Cosmologists are uncertain of the fate of a closed Universe upon reaching this superhot, superdense, infinitely small state, known among scientists as a singularity. The Universe might just end. Or it might bounce — into another cycle of expansion and contraction. The mathematics of singularities have frankly not yet been fathomed. This ultimate state of matter poses one of the hardest problems in all of science. It's perhaps fair to say that astrophysi-

cists, for the most part, are experimentally and theoretically ignorant of the physics of singularities.

Frontier research seeks an understanding of the nature of such
a singular state of matter. With both the density and temperature
increasing as the contraction nears completion, pressure — the
product of density and temperature — must increase phenomenally. The question as yet unanswered is, Will the Universe just
end as a final minuscule speck, or will this pressure be sufficient to
overwhelm the relentless pull of gravity, thereby pushing the Universe back out into another cycle of expansion and contraction? In
other words, will a closed Universe bounce?

Obviously, there is a certain amount of philosophical beauty
incorporated within such a model of an "oscillating Universe."
There is no need for a unique, once-and-for-all-time explosion —
no need for a *big* bang. Nor is there a need for a definite beginning
or a definite end. The oscillating model merely goes through phases
— an infinite number of them — each initiated by a separate explosion or bang. Indeed, in an oscillating Universe, there are many
"bangs," each expansion a "day," each contraction a "night." But
none of these bangs is unique, none of the origins any more significant than any other. Oscillation avoids the potential philosophical
hang-up of wondering about what preceded a unique big bang of
a one-cycle closed Universe or of an open Universe.

Should the oscillating model be valid, then we need not trouble
ourselves with the concept of "existence" before the beginning of
time. In this model, there is no beginning of time. Such a Universe
has always existed; it always will exist.

The above models of the Universe stipulate evolution as their
guiding principle. They are derivable from Einstein's general
theory of relativity, and they are favored in one form or another
by the majority of contemporary cosmologists. However, several
other Universe models have been proposed over the years. Most
of them do not follow directly from relativity; some don't even call
for change with time, or embrace evolution as their central concept. It's worth considering one of the more prominent ones, for

until recently it was favored by some segments of the scientific community.

The "steady-state" model stipulates not only that the Universe appears roughly the same to all observers, but also that such a Universe appears roughly the same to all observers *through all time*. Its fundamental tenet is embodied within what is sometimes called the *perfect* Cosmological Principle: To any observer at any time, the physical state of the Universe is the same. In other words, the average density of the Universe remains constant for all time. It holds steady.

Initial motivation for a steady-state model was based as much on philosophy as on science. The oscillating Universe aside, many scientists and philosophers were (and still are) unwilling to concede that nothing could have existed prior to the unique big bang. It's a most tricky question indeed to ask what preceded the origin of the Universe. What existed prior to the single, unique, big bang? Why was there a big bang? What or who initiated it? These are questions unaddressable within the realm of contemporary science. When there are no data, the scientific method becomes a useless technique. Philosophies, religions, and cults of all sorts can offer hypotheses to the nth degree, but science remains mute. The steady-state model avoids these questions, as does the oscillating model. For each, there is no beginning, and there is no end. The Universe just *is* for all time.

Steady-state cosmologists concede that the Universe is expanding, for they are unable to refute the observational fact that galaxies are receding. Steady-statists nonetheless demand that the bulk view of the Universe — the average density of matter — remains constant forever. Accordingly, since the recession of the galaxies irrevocably demonstrates that the average distance among galaxies must be increasing, the steady-state model requires the appearance of additional matter. Otherwise, with the galaxies separating, the average density would inevitably decrease. As odd as it may seem, the steady-statists propose that this matter is created from nothing. By means of the infusion of newly made matter in this way, the average density of everything in the Universe can indeed remain constant for all time. Despite the recession of the galaxies, the creation of additional galaxies in just the right amount can keep constant the number of galaxies per unit volume, thus preserving the same Universal density forever.

The most pronounced problem with the steady-state model is that it fails to specify how the additional matter is created. Nor does it specify where. Some researchers theorize its injection out beyond the galaxies in intergalactic space, whereas others prefer creation within the centers of galaxies. Not much new matter is required to offset the natural thinning as the galaxies speed apart. Creation of a single hydrogen atom in a volume equivalent to that of the Houston Astrodome every few years would do it. Unfortunately, the sudden appearance of such a minute quantity of matter, either inside or outside of galaxies, is currently quite impossible to detect, and therefore to test.

Regardless of where matter is created, the real quandary is about how it's created. The appearance of new matter from nothing violates one of the most cherished concepts of contemporary science — the conservation of matter and energy. A heavily embraced principle of modern physics maintains that the sum of all matter and all energy is constant throughout the Universe. Matter can indeed be created from energy (and energy from matter), but it's very hard to understand how that matter could be suddenly fashioned from nothing at all.

The grand puzzle of the steady-state model, then, is the process of material creation. Nonetheless, the lure of a Universe that always has existed and always will exist is strong, for it provides a way to skirt the need for a unique big bang and all the other sticky questions associated with the very start of an evolving Universe. All things considered, the big bang hypothesis is as troubling for a steady-state cosmologist to swallow as is this continual-creation hypothesis for an evolutionary cosmologist. At any rate, mental hang-ups aside, we'll now note how recent observational research has virtually eliminated the steady-state hypothesis as a feasible model of the Universe.

∞

How can we distinguish among these various possible models of the Universe? Well, the steady-state model can be ruled out almost unequivocally for at least two reasons. First, the distribution of galaxies does not appear to be uniform throughout space. Active galaxies at great distances from Earth far outnumber those nearby. Had we lived within an active galaxy (in earlier epochs of the Universe when these objects were presumably the dominant astronomical objects), then our view would have been filled with other active galaxies — many more than now surround our vantage point on Earth. The perfect Cosmological Principle is clearly violated: The large-scale view of the Universe was not the same eons ago as it is now.

A second argument against the steady-state model is the serendipitous result of experimentation. Observations made with radio telescopes always yield a signal regardless of the time of day or night. Unlike optical observations for which there is sometimes a complete void of light toward the dark and obscured regions of space, radio observations never fail to detect some radiation. Sometimes the radio signal is strong, especially when the telescope is aimed toward an obvious source of radio emission. Other times it's weaker, particularly in regions devoid of all known radio sources. Yet, whenever the accumulated emission from all the known radio sources and from all the instrumental noise is accounted for, there still remains a minute radio signal — sort of a weak hiss, not unlike static on an AM radio. Never diminishing or intensifying, this weak signal is detectable at any time of the day, any day of the year, year after year. What's more, it's equally intense in any direction of the sky — that is, it's isotropic. The whole Universe is apparently awash in this low-level radiation.

This omnipresent radio signal was accidentally discovered little more than a decade ago during an experiment designed to improve America's satellite communications system. In their data, scientists unexpectedly noted the bothersome radio hiss that just wouldn't go away. Unaware that they had detected cosmologically significant radiation, they sought many different origins for the excess emission, including atmospheric storms, ground interference, antenna imperfections, even pigeon droppings deposited inside the radio telescope. Subsequent conversations with theorists enlightened the experimentalists as to the static's most probable origin. That origin is the entire Universe itself.

The weak, isotropic radio radiation is widely interpreted as a veritable "fossil" of the primeval explosion that commenced the Universal expansion long ago. This relic hiss is often referred to as the cosmic background radiation. Its existence is completely consistent with any of the evolutionary models of the Universe, though there is no role for it in the steady-state model.

The cosmic background radiation is presumed to be a remnant of the exceedingly hot phase of the early Universe — a Universe that has since cooled during the past fifteen billion years or so. Regardless of whether the initial explosion was a unique big bang producing an open and infinite Universe, or one of several smaller bangs leading to a closed, finite, and oscillating Universe, the primeval, hot, dense matter must have emitted thermal radiation. All objects having any heat emit radiation; a very hot piece of metal, for instance, glows with a red- or white-hot light, whereas less-hot metal feels warm to the touch and emits less energetic infrared or radio radiation. In its fiery beginnings, the Universe almost certainly released extremely energetic gamma-ray radiation. But with time, the Universe expanded, thinned, and cooled. Consequently, the emitted radiation is predicted to have steadily shifted from the gamma-ray variety normally associated with superhot matter, down

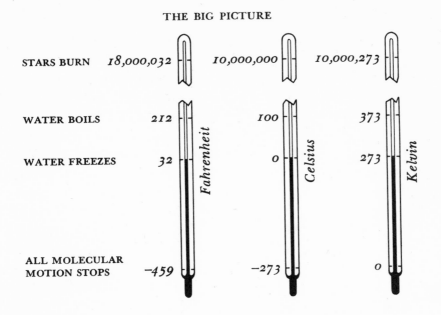

	Fahrenheit	Celsius	Kelvin
STARS BURN	*18,000,032*	*10,000,000*	*10,000,273*
WATER BOILS	*212*	*100*	*373*
WATER FREEZES	*32*	*0*	*273*
ALL MOLECULAR MOTION STOPS	*-459*	*-273*	*0*

through the less energetic ultraviolet, visible, and infrared, and eventually to the radio variety normally associated with relatively cool matter.

Evolutionary models predict that some fifteen billion years after the primeval explosion, the average temperature of the Universe — the relic of this big bang — should be quite cold, in fact no more than about −270 degrees Celsius. That's far below the 0 degrees Celsius temperature at which water freezes, and only a few degrees above the absolutely coldest value at which all atomic and molecular motion ceases. On the scientific scale, −270 degrees Celsius is equivalent to 3 degrees Kelvin. To confirm the theory, researchers have carefully measured the intensity of the weak isotropic radio signal at a variety of frequencies. The best-fit solution for all the data acquired during the past decade is indeed consistent with a Universal temperature of approximately −270 degrees Celsius, or 3 degrees Kelvin. Furthermore, this oldest fossil really does seem to pervade the whole Universe, including Earth, the building and the room in which you now find yourself. The amount of cosmic radiation absorbed at any one time, however, is minuscule, totaling less than a millionth of the power emitted by a hundred-watt light bulb.

Existence of the cosmic background radiation, together with the spread of galaxies in space, discredits the steady-state hypothesis as a feasible model of the Universe. Clearly, the Universe has changed with time; it hasn't been steady at all. The choice of correct Universe model must then be made from among the evolutionary ones. Other data must be acquired to sift through each of them.

The most straightforward way to distinguish between the open and closed models requires an estimate of the average density of matter in the Universe. More than anything else, density is what differentiates the closed model, which has enough matter to halt the expansion before it reaches infinity, from the open model, where there simply isn't enough to bring it all back.

It would be foolhardy to try to inventory all the matter in the Universe. Authors don't try to count all the words in a manuscript, but rather make an estimate by counting the words on a single page, and multiplying by the number of pages. Likewise, astrophysicists can make an estimate of the amount of matter within a certain volume of space, and then extrapolate that amount to include the whole Universe. This is tantamount to estimating the mass density, for density is nothing more than mass per unit volume.

The precise density of matter needed to halt the expansion just as the outer limits of the Universe reach infinity can be computed theoretically. For today's thinned-out Universe, the answer is some million million million million million times less than one gram per cubic centimeter. (A cubic centimeter is just about the volume contained within a small sewing thimble, while a gram is about one five-hundredth of a pound. A thimbleful of water would have a mass of about one gram, the density of that liquid being approximately one gram per cubic centimeter.) This extraordinarily small density amounts to a few hydrogen atoms within a volume the size of a typical household closet. That's terribly tenuous; in fact, many orders of magnitude thinner than the best vacuum attainable in terrestrial physics laboratories. But remember, this is an average density of the entire Universe — lumping galaxies, where the matter is, together with intergalactic space, where little if any matter resides.

If the actual density of the Universe is less than this theoretically computed value, then the Universe is destined to expand forever, the hallmark of an open, infinite, hyperbolic model. If, on the other hand, the actual density exceeds this value, the Universe will someday stop expanding and start contracting, the fate of a closed, finite, elliptical model.

Theory aside, how can we determine the actual density of matter in the Universe? At face value, it would seem simple. Just measure the average mass of each of the galaxies residing within any parcel of space, estimate the volume of that space, and calculate the total density. Having done this, researchers often find about ten times less density than the amount needed to halt the expansion of the Universe. As far as can be determined, this calculation is independent of whether the chosen region contains only a few galaxies or a rich cluster of galaxies; the resulting density is pretty much the same, within a factor of two or three.

Galaxy-counting exercises of this sort imply that the Universe is open, meaning that it originated from a unique big bang and will expand forevermore. Such a Universe has no end, but it definitely had a beginning.

An important caveat deserves mention here. All the matter in the Universe may not be housed exclusively within the brightly visible galaxies. Recent observations suggest that some invisible matter does exist out beyond each of the galaxies. The extent and amount of it is presently unclear, but if more than ten times as much additional matter resides outside the galaxies as in the galaxies themselves, then the Universal density would correspondingly increase by a factor of more than ten. Reservoirs of matter skirting the galaxies could then reverse the solution to this first cosmological test, forecasting a closed Universe possibly having an end as well as a beginning. Whether such a Universe originates from a unique big bang prior to which there existed nothing, or whether such a Universe ends without bouncing, cannot be addressed by this test.

The value of the Universal density determined by this galaxy-counting method is thus really quite uncertain at the present time. It cannot unambiguously distinguish between the open and closed models, though at face value it favors an open Universe that will recede evermore.

There is another observational test that seeks to determine the ultimate fate of the Universe. Like the method above, this second test attempts to estimate the average mass density of the Universe. It essentially relies on the fact that each and every piece of matter gravitationally pulls on all other pieces of matter. Specifically, this second test addresses the question: At what rate is matter everywhere causing the Universal expansion to slow down? Put another way, how fast is the Universe decelerating?

If the Universe began in an explosive bang, it must have expanded rapidly at first, gradually growing more sluggish. The expansion of anything — an ordinary bomb, an atmospheric thunderclap, whatever — is always greater at the moment of explosion than at some later time. Hence, since looking out into space is equivalent to looking back into time, the recession of the galaxies should be larger for the faraway galaxies, and a good deal smaller for those nearby.

Operationally, cosmologists try to detect any change in the recession velocity of our neighboring galaxies compared to those far away. This change is expected to be greater for the finite, closed model of the Universe, since the large amount of matter necessary to contract the Universe would have considerably decelerated the Universal expansion over the course of fifteen billion years. The infinite, open model is predicted to show smaller change in the galaxy recessional velocities, for in this case the deceleration of the Universe is less.

What do the data indicate? Is there any evidence for a faster recessional velocity among the more distant galaxies? In a nutshell, the most distant galaxies do seem to show a substantially greater recessional velocity than those nearby. The accuracy of these observations is rather poor, however. The most distant galaxies are obviously faint, and observations of them are notoriously hard to make. Nonetheless, this second cosmological test suggests that the Universe is closed and finite. It contradicts the first test, unless substantial amounts of matter really do exist far beyond the galaxies.

Whether we live in an open or closed Universe, then, is currently unknown. The bottom line clearly suggests an evolutionary Universe, but its ultimate destiny remains concealed. Many cosmologists are inclined to say (and have been saying for the past couple of decades) that we should expect a definite answer within a few years. This is perhaps overoptimistic, for the final solution requires the agreement of three often disparate groups of human beings. First, there are the theoreticians, those imaginative minds who invent the model Universes. They try to determine what the Universe is supposed to be like. Second, there are the experimentalists, constantly testing the theories, all the while extending their observations to more distant realms within the real Universe. They try to determine what the Universe really is like. And third, there are the skeptics, who regard the models of the first group as mere speculation, and the results of the second group as overinterpretation of data without regard for observational errors. In the end, all three attitudes are desirable, for only by their coordination and counteraction may we ever hope to approach the truth.

EPOCH ONE

PARTICLES

Order by Chance?

billions of years ago

WHAT WAS IT LIKE at the start of the Universe? Precisely what happened at the origin of time? Can anything concrete be said about the very origin itself, or about the prevailing conditions during the first few moments of the Universe?

These are surely fundamental questions. They are hard questions. Yet they are among the questions that perhaps every human who has ever lived has contemplated at one time or another. Now, after more than ten thousand years of organized civilization, twentieth-century science seems poised to provide some insight regarding the ultimate origin of all things.

The solutions astrophysicists have developed should be considered qualified and provisional. Times long past are times long gone. It's difficult to be precise about that which cannot be observed directly. Nonetheless, models can be constructed — mathematical sketches based on a lode of relevant data dictating the shape and structure of our Universe. These models grant us some inkling of what the Universe was like more than ten billion years ago.

∞

To appreciate the earliest epochs of the Universe, we must be willing to think deeply about times long ago. We must strive to

38

imagine what it was like long before Earth and Sun were created — even before any planet or star existed. Some people have difficulty mentally visualizing such truly ancient times. Fortunately, there's a trick that can help us comprehend the earliest moments of the Universe.

Physicists are principally charged with the application of the laws of nature to the present state of something in order to predict its future. In the case of the big picture, that "something" is the whole Universe. Hence, if we find it hard to mentally reverse time to appreciate the earliest epochs of the Universe, we can instead take advantage of the natural symmetry of a model Universe that will eventually contract, and attempt to predict the physical events destined to occur as a closed Universe nears the final phase of its collapse. This procedure is valid only because the mathematics describing a contracting Universe is a mirror image of the mathematics describing an expanding Universe. The events that *will* occur just prior to the total collapse of a contracting Universe mimic the events that *already* happened just after the Universe began expanding. Using the laws of physics in order to predict the final events of a contracting Universe, we can gain some understanding of the early aftermath when the Universe banged approximately fifteen billion years ago. Simply put: Conditions near the death of the Universe may uncannily resemble those just after its birth.

Even if the Universe is not closed, and will never collapse to a singularity, astrophysicists can use the closed model to understand theoretically the highlights of the earliest epochs of either a closed or an open evolutionary Universe.

Numerical experiments are required to crank out these Universe models. The experiments are essentially number-crunching exercises, utilizing a mathematical knowledge of the laws of physics and a large computer. The calculations are complex, incorporating the essentials of all that we know about the Universe. The objective is to determine the average density and temperature for the whole Universe at any point in time. The input numbers can be

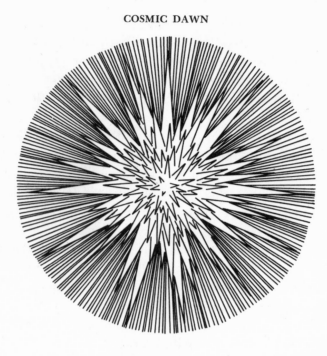

varied, in order to match the present state of the model Universe with that of the real Universe.

Most numerical experiments suggest that in the beginning, there was chaos! It's not really possible to inquire about what might have happened at the exact moment of the bang — precisely zero time. But some theorists argue that the physical conditions can be specified for some extraordinarily short time after the bang. For example, the laws of physics specify that a Universe younger than a trillionth of a trillionth of a second (i.e., 10^{-24} seconds) would have had an average density greater than a trillion trillion trillion trillion (i.e., 10^{48}) grams per cubic centimeter, and an average temperature greater than a trillion degrees Celsius. By way of comparison, the average density of water is one gram per cubic centimeter, of lead about five grams per cubic centimeter, and of atomic nuclei a mere trillion grams per cubic centimeter. The present average density of everything in the Universe is something like a million trillion trillion times *less* than that of water (or about 10^{-30} grams per cubic centimeter); this is the average density of *everything* — galaxies, stars, people, as well as mostly empty space. Similarly, water freezes at 0 degrees Celsius, and boils at 100 degrees Celsius, while the average temperature at the surface of an ordinary star is several thousand degrees Celsius. The present average temperature of everything in the Universe is only a few degrees above the temperature at which all atomic and molecular motion ceases, some -270 degrees Celsius (or 3 degrees Kelvin).

As for the time, it's virtually impossible to appreciate such youth; 10^{-24} seconds is the amount of time light needs to cross a proton. It's quicker than a flash! Such minute fractions of time are as incomprehensible as the large densities and temperatures characterizing the early Universe. Yet these are the conditions predicted by the laws of physics as a contracting Universe inexorably speeds toward its demise. They are thus, through the above symmetry arguments, the conditions prevailing at the earliest moments after the birth of the Universe.

The composition of the Universe at this time was indescribable. Surely, much energy existed, along with exotic elementary particles of all types, but beyond that science can say little. The dominant action at the start of the particle epoch must have been simply unimaginable.

By most accounts, then, the Universe originated with the explosion of an unbelievably hot and dense something. Precisely what something, we cannot say. And why it exploded, we really don't know. Why the Universe suddenly began expanding more than ten billion years ago is a most intractable query — so formidable that scientists are currently unaware even of how to go about formulating it.

In the broadest sense, there are *what* questions, *how* questions, and *why* questions. Unraveling microscopic and macroscopic matter with biochemical microscopes and astrophysical telescopes, scientists are able to address fairly well *what* there is in the Universe — from atoms to galaxies. Armed with this detailed inventory of matter, we are then able to address the origin and evolution of that matter, in other words, *how* it got there and *how* it changed to its present state.

To inquire about the nature of the very beginning, however, requires us to address *why* questions, the most fundamental of all being Why is there a Universe? Scientists simply don't know how to address why questions. They are outside the present formulation of modern science. There is no known procedure — not even the scientific method — for investigating why the laws of physics and biology are as they are. We'll probably never know the answer. Nor do we know, or have any prospect of ever knowing, why there is a Universe.

The basic difficulty in attempting to discover the nature of what might have preceded the very start of the Universe is simple: There are no data. None whatsoever. Sure, some people have hypotheses, but these hypotheses are not based upon data. Furthermore, there is currently no way that scientists can experimentally test these various hypotheses.

This is not a criticism of humans who wonder about the start of the Universe. Some scientists often wonder how they might devise experiments to gather pre-Universal data. However, as things stand now, queries about the nature of whatever existed before the bang amount to inquiries, not about the origin of the Universe, but about the origin of the origin.

In what follows, we necessarily confine our discussion to events that have occurred since the beginning of the Universe, regardless of why the Universe did originate. Indeed, cosmic evolution constitutes a broad synthesis of the whats and hows.

Within a microsecond of its being, the Universe was filled with a whole mélange of microscopic particles of matter. This was a straight materialization — a creation — of matter from the energy of the primeval bang. No black magic is involved here, just a well-known and oft-studied nuclear process whereby elementary building blocks of matter result from clashes among packets of energetic radiation.

Called hadrons, the heavy elementary particles such as protons and neutrons were the most abundant type of matter at the time. Such particles must have existed as free unbound entities, considering the astronomical temperatures prevalent in the Universe well within its first second of existence. It was just too hot for these particles to assemble into anything more substantive. Hadrons unquestionably collided and interacted with one another and with other types of elementary particles, for the density was also extremely great. According to theoretical models, the dominant action at this time was the self-annihilation of hadrons into radiation, thus creating a brilliant fireball. Lacking a good understanding of elementary particles, physicists know little about this mystifying period in the history of the Universe.

The basic stuff of the Universe continued to fly apart rapidly, cooling and thinning all the while. About a millisecond after the bang, the superhot and superdense conditions suitable for hadron creation had nearly subsided, thus allowing the less abundant, lighter elementary particles such as electrons and neutrons to predominate. This began another process of materialization whereby a whole new class of lightweight particles, called leptons, were fashioned from energy under an average density of ten billion grams per cubic centimeter and a temperature of about ten billion degrees Celsius. These physical conditions were still excessive by terrestrial standards, but they had diminished considerably compared to the chaotically dense and hot conditions extant a fraction of a second earlier. Indeed, it's important to realize that once the Universe began expanding, it did so very rapidly, unhesitatingly cooling and dispersing its contents. By the time the first second had elapsed, leptons were self-annihilating into radiation, much as had the hadrons earlier. The radiative fireball of this cosmic bomb was still fueled with light and other types of radiation.

The radiation density greatly exceeded the matter density in the first few minutes. Not only did the waves of radiation far outnumber the particles of matter, but also most of the energy in the Universe was in the form of radiation, not matter. No sooner did elementary particles try to coagulate than the fierce radiation destroyed them, thus precluding the existence of the simple type of matter we now call atoms. For this reason, the earliest portion of the particle epoch is sometimes referred to as the Radiation Era. Whatever matter managed to exist did so as an inconspicuous precipitate suspended in a sea of dense, brilliant radiation.

As time elapsed, change continued. A few hundred years after the bang, the density had decreased to a value of about a billionth of a gram per cubic centimeter, while the average temperature had decreased to about a million degrees Celsius — values not terribly different from those in the atmospheres of stars today. A principal feature of this latter portion of the particle epoch was the steady waning of the original fireball; all the annihilations of hadrons and leptons had virtually ceased. Even as the fireball faltered, though, a most important transformation began.

The first few centuries of the Universe saw radiation reign supreme over matter. All space was absolutely flooded with light, X rays, and gamma rays. As the Universe expanded in time, however, the radiation density decreased faster than the matter density. The early fog of blinding light gradually thinned, thus diminishing the early dominance of radiation. Sometime during the interval from the first few minutes to a million years after the bang — the exact moment cannot yet be pinned down any better — the charged elementary particles of matter began to assemble. Their own electromagnetic forces pulled them together, sporadically at first, then more frequently. Radiation could no longer break them apart as quickly as they combined. In effect, the authority of radiation had subsided as matter gradually became neutralized, a physical state over which radiation has little leverage. Matter had, in a sense, overthrown the cosmic fireball while emerging as the principal constituent of the Universe. To denote this major turn of events, the latter portion of the particle epoch and all the remaining epochs known to have since occurred are collectively called the Matter Era.

The emergence of matter from radiation is the first of two critically important transformations in the history of the Universe.

So the onset of the Matter Era saw the creation of atoms. The influence of radiation had grown so weak that it could no longer prohibit the attachment of the lepton and hadron elementary particles that had survived annihilation. Hydrogen atoms were the first type of element to form, requiring only that single negatively charged electrons be electromagnetically attracted to single positively charged protons. In this way, copious amounts of hydrogen were synthesized in the early Universe.

Hydrogen is the common ancestral element of all things.

Hydrogen was not the only kind of atom fashioned during the particle epoch. Before all the electrons and protons were swept up into hydrogen, atoms of the second simplest element, helium, began to materialize.

Heavy elements inevitably originate when two or more light elements fuse together. They do so by means of a two-step process. First, a heavy nucleus of an atom is created whenever lighter ones collide violently. Second, the newly formed positively charged nucleus then attracts a requisite number of negatively charged electrons, thereby yielding a neutral, albeit heavier, atom.

In the case of helium creation, a temperature of at least ten million degrees Celsius is required to thrust two hydrogen nuclei (protons) together; each boasts a positive charge, and at lower temperatures they would simply repel one another like identical poles of magnets. This minimum temperature ensures that the hydrogen nuclei collide with ample vigor to pierce the natural electromagnetic barrier that prevents them from interacting under ordinary circumstances. For a split second, the colliding particles enter the extremely small operating range of the powerful nuclear force. Once within about a trillionth of a centimeter of one another, the two hydrogen nuclei no longer repel. Instead, the attractive nuclear force seizes control, slamming them together ferociously, and uniting them instantaneously into a heavier nucleus. Exactly the same process is occurring right now in the hearts of stars everywhere. And it's the same process that humans have made operational, though on a much smaller and uncontrolled scale, in the form of modern thermonuclear weapons, especially the hydrogen bomb.

In the late stages of the particle epoch, the physical conditions were ripe for the creation of helium nuclei from protons of the primordial fireball. Shortly thereafter, pairs of electrons were electromagnetically attracted to each helium nucleus, thus fabricating neutral helium atoms. Given the rate at which most models suggest the Universe expanded and cooled, only so much of the hydrogen could have been transformed into helium, leaving about ten hydrogen atoms for every one helium atom.

By contrast, elements heavier than helium could not have been appreciably produced in the early Universe. Elements composing this page you are reading, the air we breathe, and the coins in our pockets were not created in the aftermath of the initial explosion. Fusion of heavier elements, such as carbon, nitrogen, oxygen, iron, and uranium, requires temperatures much higher than ten million degrees Celsius. Such syntheses also require lots of helium atoms. The basic trouble here is that, even though the helium atom production was in high gear during the latter portion of the particle epoch, the average temperature was quickly falling. The Universe after all was rapidly expanding, the result of its primeval explosion. Theoretical calculations suggest that, by the time there were sufficient helium atoms to interact with one another to manufacture some of the heavier elements, the cosmic temperature had

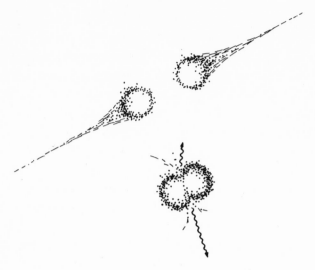

fallen below the threshold value required for the mutual pene-
tration of doubly charged helium nuclei. That threshold value is
a hundred million degrees Celsius, for it takes great violence for
doubly charged nuclei to collide, stick, and fuse. The Universe
just wasn't hot enough anymore to permit this.

Contrary to the progressive cooling and thinning of the early
Universe, the compact matter within stars, not yet arisen at that
time, is perfectly suited for the generation of hotter temperatures,
greater densities, more brutal collisions, and thus heavier elements.
The guts of stars are indeed where the heavies were eventually
created, albeit long after the particle epoch. It is where they are
still being created today.

∞

An atom of ordinary matter is an invisible microscopic entity
made of a positively charged heavyweight nucleus, usually several
protons and neutrons, surrounded by one or more negatively
charged lightweight electrons. All atoms found on Earth and the
Moon have this common structure. Furthermore, radiation re-

ceived from extraterrestrial objects, near and far, is consistent with this same basic structure for all atoms everywhere.

Some theorists wonder if there might be other kinds of atoms — not just different elements, but atoms that are built with fundamental differences from the ones we know on Earth. Why, for instance, should the heavy nuclei always have a positive charge, relegating the negative charge to only the lightweight electrons? They argue that the Universe would be more philosophically pleasing if its basic building blocks had more symmetry in their charge and mass. In other words, perhaps the Universe is also endowed with atoms made of negatively charged nuclei around which orbit positively charged particles. Experimentalists in fact discovered, around the mid-twentieth century, several lightweight particles having a positive charge, and we now know of several heavyweight particles having a negative charge. These so-called antimatter particles are identical to ordinary matter particles in every way except charge. A particle called a positron, for example, is just like an electron, but with a positive charge. These same experiments also showed that when a matter particle and its antimatter opposite collide, the result is an explosion that releases pure energy of the lethal gamma-ray variety.

The reverse phenomenon can occur as well. Provided the temperature is extraordinarily great (in the range of billions of degrees Celsius), collisions among packets of gamma radiation can produce pairs of elementary particles, for instance, a matter electron and an antimatter positron. This sort of "materialization" of matter and antimatter from energy still obeys the fundamental laws of physics; in this case, one (energy, or E) simply converts into the other (mass, or m), the conversion obeying the famous relationship $E = mc^2$, where c symbolizes the velocity of light.

This and other kinds of conversion from energy to mass is precisely what theoretical models suggest happened in the earliest moments of the Universe. Yet we don't observe much antimatter around us. Earth, the other planets, and the Sun all appear to be composed of ordinary matter. Exceptions include some particles produced in nuclear reactions known to be churning away inside stars, a small fraction of the baffling cosmic rays showering Earth each day, and microscopic fragments created during fleeting moments when elementary particles collide in nuclear laboratories here on Earth. Still, virtually all the mass in the Solar System seems to be of the matter variety with little trace of naturally occurring antimatter. If matter and antimatter were created in equal amounts from primordial energy at the very start of the Universe. then where has all the antimatter gone?

It's important to realize that antimatter does not imply antigravity. Particles of antimatter gravitationally attract one another just as do two or more particles of matter. The only property distinguishing matter from antimatter is charge; the mass of every matter particle is identical to that of its antimatter opposite, and hence gravity invariably pulls while never pushing. To be sure, there is no such thing as "antigravity" known anywhere in the Universe.

Nothing, in principle, prohibits elementary particles of antimatter from combining into large clumps. Antihydrogen, antioxygen, anticarbon, and numerous other antiatoms could conceivably form antiplanets, antistars, and antigalaxies. The fact that we are unaware of such large clumps of antimatter does not preclude their existence. Since atoms of antimatter emit and absorb precisely the same type of photons as do atoms of ordinary matter, there is no way of determining if, for example, a distant star is composed primarily of matter or antimatter. Spectroscopic observations of a cluster of antimatter would be identical to those made of a clump of matter. Accordingly, the nearby Alpha Centauri star system or the Andromeda Galaxy as well as numerous other stars and galaxies *could* be composed of antimatter.

Despite the fact that our Solar System is made primarily if not totally of matter, there may well exist large pockets of antimatter elsewhere in the Universe. Provided coagulations of matter remain separated from those of antimatter, then they can coexist without difficulty. As to where the primeval antimatter might be, we can conjecture that it's wrapped up in large, distinct assemblages far outside our Solar System. Should similar matter and antimatter

49

objects venture too close together, however, they would mutually annihilate. Consequently, if our civilization ever attains an ability to travel beyond our Solar System, it will be important to dispatch automated probes before visiting alien worlds. Should such an unmanned spacecraft suddenly evaporate in a puff of gamma radiation, it would be in our best interest to visit elsewhere.

Be sure to recognize that there is at present no experimental evidence for the existence of antimatter objects beyond Earth. They are only theoretically inferred on the basis of the symmetry of matter and antimatter thought to have been created from energy in the earliest moments of the Universe.

∞

By the end of the particle epoch, the Universe had evolved dramatically. The spectacularly bright fireball associated with the origin of all things had subsided. The physical conditions of temperature and density that guide all changes in the Universe had themselves undergone extraordinary change. Matter had wrested firm control from radiation, heralding a whole new era. Atoms, exclusively hydrogen and helium, had been synthesized.

Major events in the Universe would thereafter be less frequent. Change continued, though at a more relaxed pace. Important transactions between matter and radiation may well have occurred posthaste immediately after the bang, and especially in the first few minutes of the Universe. But these interactions eventually lessened, becoming few and far between by the end of the particle epoch. The average density decreased enormously throughout this initial epoch, plummeting below a billionth of a billionth of a gram per cubic centimeter before the epoch ended — roughly a million years after the Universe began. The average temperature of the entire Universe had also diminished by this time to a relatively cool thousand degrees Celsius.

With time, the Universe had grown thinner, colder, and darker. It was destined to evolve much more slowly in the later epochs, but it evolved nonetheless. The average physical conditions were on their way to becoming a billion times still less dense and a thousand times still less hot — tenuous and frigid conditions now present more than ten billion years after the bang — the fossilized grandeur of an ancient and glorious era.

The history of the early Universe presented here represents the prevailing view among cosmologists. However, all do not share a consensus concerning events prior to the first second of existence. The farther we attempt to probe in the past, the more uncertain our statements become. Accordingly, the temperature and density in the first instant of the Universe are quite obscure, mainly because their values depend upon incompletely understood interactions among the heavy elementary particles.

Subject to the intricacies of the theoretical model chosen, the dimensions of the events during the first few seconds can differ by several orders of magnitude. It's not surprising, for the earliest moments of the Universe are gone with the expansion, forever lost to evolution. We can fathom the vastly ancient realms of nature only indirectly, aided by crutches of abstract formulae and logical symbols. Yet what we find, in virtually all models, is an early Universe considered to have been exceedingly hot and dense, growing cooler and thinner with the advance of time, and gradually promoting the successive formation of galaxies, stars, planets, and life.

EPOCH TWO

GALAXIES

Hierarchy of Structures

billions of years ago

DESCENDANTS OF OUR CIVILIZATION may never become advanced enough to journey far away from our Milky Way Galaxy — far enough to look back and witness its full grandeur. The big picture of our swarm of starlight floating proud and silent in the near void of space may forever elude us. Yet, from our Earth-based vantage point at one edge of our Galaxy, it is possible to examine the full extent of other colossal star systems far beyond our own Milky Way.

Deep space contains myriads of objects looking strangely unlike stars. Many have a fuzzy, lens-shaped appearance, often resembling a disk rather than the clear, bright, spherical image usually associated with stars. The eighteenth-century German philosopher Immanuel Kant regarded them as individual "island universes" well beyond the confines of our Milky Way Galaxy. To label each of them a "universe" presents a clear semantics problem, but he was correct in arguing that these nonstellar patches of light reside outside our Galaxy.

Large telescopes have since revealed these blurry and distant beacons to be entire galaxies, each measuring some hundred thousand light-years across. Replete with literally hundreds of billions of stars bound loosely by gravity, each galaxy harbors more stars than people who have ever lived on Earth. Silently and majestically,

52

galaxies twirl in the vast reaches of the Universe — huge pinwheels of radiation, matter, and perhaps life — simultaneously imparting to us a feeling for the immensity of the Universe and for the mediocrity of our own position in it.

Objects identified as galaxies in photographs usually have spiral shapes pretty much like our Milky Way or the neighboring Andromeda Galaxy. They have a central bulge from which emanate thin spiral regions, or "arms," chock-full of stars. The apparent prevalence of spiral galaxies is just that — apparent — resulting largely from the fact that whorled patterns are most easily discerned from among the many other patches of light in the nighttime sky. In reality, galaxies have many morphologies, and the spiral one is not the most abundant type of galaxy in the Universe.

After decades of effort, inventories are now virtually complete for that part of the Universe in which we live. Most abundant are galaxies shaped like footballs, and officially called elliptical galaxies. Some are less elongated, approximating more the shape of beach balls; others resemble fat cigars. Regardless of shape, each elliptical galaxy harbors the usual hundred billion or so stars, strung across some hundred thousand light-years, though devoid of spiral arms. Elliptical galaxies seem to be simply undistinguished, though monumental, globs of stars.

Spiral arms are not the only trait that elliptical galaxies lack. Hardly any gas and dust drifts among the stars of elliptical galaxies — that is, they have little or no interstellar matter. This implies that all the elliptical galaxies are old. Stars, which normally originate from interstellar matter, apparently did so long ago, leaving no loose gas and dust for the continued formation of future generations of stars. Analyses of the radiation emitted by individual stars within elliptical galaxies further demonstrate that all of them are old. Evidently, nearly all the interstellar gas and dust was used up eons ago, thus quenching the star formation process.

The lack of gas, dust, and star formation in the elliptical galaxies contrasts sharply with the abundance and activity of interstellar matter within the second galaxy class — spiral galaxies. Here, there

are also a variety of shapes, though all of them are basically flattened disks resembling two sombreros, their brims clapped together. Some spiral galaxies have a large central bulge around which there are fairly tightly wrapped spiral arms. Others have a more open pattern of arms emanating from an intermediate-sized central region. Still others have a rather small center from which long, stringy arms protrude, sometimes making it difficult to recognize these as spirals.

Spiral galaxies are known to contain lots of gas and dust mixed throughout the vast spaces among their stars. Furthermore, observations during the past decade show that stars are still forming, most of them in the arms. Unlike the old elliptical galaxies, spiral galaxies have a good deal of vitality. This doesn't mean that the spiral galaxies are necessarily young. Rather, they are simply still rich enough to provide for continued stellar birth.

There is a third galaxy type. Called irregular galaxies, these are large coagulations of stars, gas, and dust whose visual appearance prohibits their being placed into either of the two previous cate-

ories. By and large, irregulars tend to be a bit smaller than other
types of galaxies. Some researchers refer to them as dwarf galaxies.
They often do seem to be dwarfed by the larger spiral or elliptical
galaxies near which they are usually found. In fact, there's no clear
evidence that irregular galaxies exist alone anywhere in space;
they are invariably allied with a "parent" galaxy from one of the
other two categories.

Our Milky Way Galaxy has two small companion irregular
galaxies. One is slightly larger than the other, but each is about
ten times as small as our own Milky Way system. They probably
orbit our Galaxy, just as Earth orbits the Sun, or the Moon orbits
the Earth. The motions of these irregular galaxies are slow by
human standards, however, and their orbital paths have not yet
been firmly established.

Called the Magellanic Clouds, they are named for the sixteenth-
century Spanish voyager Ferdinand Magellan, whose round-the-
world expedition first brought word of these great fuzzy patches of
light to the European civilizations in the Northern Hemisphere.
They can be viewed, however, only from locations south of Earth's
equator. Seen easily with the naked eye and looking like dimly
luminous atmospheric clouds, the Magellanic Clouds have undoubtedly served as sources of celestial wonder to residents of the
Southern Hemisphere since the dawn of civilization.

∞

Our planet Earth is finite; beyond it stretches the tenuousness of
interplanetary space. Our Solar System is finite; beyond it lies the
near-vacuum of interstellar space. And our Galaxy is finite; beyond it exists the absolute void of intergalactic space. Perhaps then
the arrangement of galaxies in space is also finite. The question
comes to mind: How are the galaxies spread through the expansive
tracts of all space? Is there some terminal point beyond which
galaxies can no longer be observed? Or do they reside everywhere,
all the way out to the very limits of the observable Universe?

A variety of techniques can be used to estimate the distances to

the galaxies. Within the nearby realm of a few million light-year
we know of about twenty galaxies. Giant spiral galaxies, such
our own Milky Way and Andromeda, are spread among mar
dwarf irregulars, such as the Magellanic Clouds. Evidently, the
twenty galaxies are clustered together as a result of their mutu
gravitational attraction — a larger version of the same natural ph
nomenon that holds stars in galaxies, planets in star systems, ar
people on Earth. In all, these twenty "local" galaxies are clustere
within a volume whose diameter is some three million light-year
Including our Milky Way, the whole bunch is known as the Loc
Group. It constitutes our neighborhood in space.

Three million light-years is a very large piece of real estate. B
don't let it confuse you. Be sure to comprehend two things her
First, recognize that we've suddenly made a large jump in spati
dimensions, from earlier considerations of the hundred-thousan
light-year size of our Milky Way Galaxy to the three-million-ligh
year size of our Local Group of galaxies. Second, recognize that th
Milky Way does not lie at the center of this cluster of galaxies. N
only is Earth not the center of our Solar System, and the Sun n
the center of our Galaxy, but our Galaxy is also not the center
the much larger Local Group. Though we may like to think so, w
are not at any special, unique, or preferred location in this ga
gantuan, perhaps infinite, Universe.

Be assured, many more than twenty galaxies reside in the Un
verse. Photographic time exposures made with large telescop
reveal thousands of galaxies within virtually any small field
view. In all, estimates suggest that there are some hundred billic
other galaxies in the observable Universe.

Interestingly enough, all the other galaxies are much farth
away than even the distant members of our local galaxy cluste
For millions of light-years beyond the edge of the Local Grou
there appears to be nothing — no galaxies, no stars, no gas ar
dust — nothing — just empty intergalactic space.

Strive to imagine the recesses of deep space beyond the Loc
Group. Passing through a seemingly interminable void, we con

>on a few more galaxies scattered here and there. But not until
e reach a distance of some fifty million light-years away do we
icounter a rich galaxy cluster, an unmistakable volume of space
imming with galaxies. Containing not just twenty galaxies like
ir own Local Group, this so-called Virgo Cluster harbors about a
ousand galaxies. Can you imagine a thousand individual galaxies
l clustered in a swarm, each one housing about a hundred billion
irs? No wonder humans — even astrophysicists — have trouble
ntemplating the immensity of matter, space, and time in the
niverse.

Galaxy clusters like these exist everywhere in the Universe.
hey are not products of our imagination. Their presence is fact.
i much the same way that galaxies are collections of stars, galaxy
usters are collections of galaxies. They constitute a most lofty
ing in the hierarchy of material coagulations within the Universe
- particles, atoms, molecules, dust, planets, stars, galaxies, and
ow galaxy clusters.

When contemplating rich galaxy clusters, such as Virgo with
s thousand or so members, it's hard to avoid the impression that
ere must be galactic traffic jams. Just as there are collisions among
oms clustered in an enclosed box or hockey players in a limited
nk, the random motions of galaxies within a galaxy cluster could
inceivably induce phenomenal collisions among these giant coag-
ations of matter.

Galaxies do indeed collide. A good deal of observational evi-
ence proves that they do so, and quite often. Numerous celestial
iotographs show what appear to be two or more galaxies in the
rocess of interacting. While many galaxies seem to lie along the
me line of sight, though they are really very much separated
om each other, others are actually near one another in space.
'hether they are genuinely colliding or only experiencing a close
icounter cannot easily be determined, for appreciable motion of
distant galaxy typically takes millions of years.

At first thought, a collision among giant galaxies would be ex-
icted to create a mind-boggling crunching of matter, complete

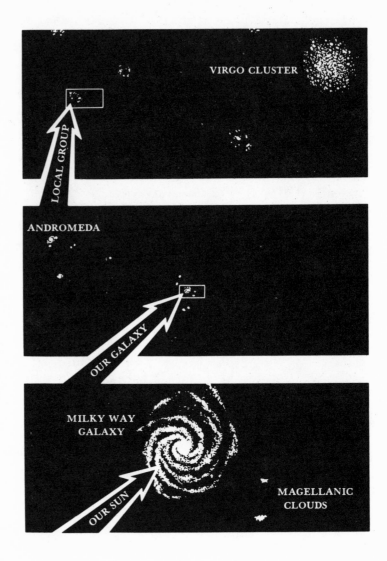

with spectacular explosions and superlative fireworks. Surprisingly, that doesn't happen at all. Such collisions, in fact, are rather quiescent: The stars in each galaxy more or less just slide past one another. Why? Because stars hardly ever collide; they are, after all, small objects by cosmic standards. While we have plenty of direct photographic evidence for galaxy collisions, no one has ever succeeded in witnessing or photographing a collision between two stars — not even in our own Milky Way.

We can explain this oddity by noting that galaxies within a typical cluster are bunched fairly densely. The distance between adjacent galaxies in the same cluster averages about a million light-years, which is only about ten times as great as the size of a typical galaxy. This doesn't really give them much room to roam around without bumping into one another. By contrast, stars inside any galaxy are spread considerably thinner. The average distance between stars in a galaxy is about five light-years, millions of times as great as the size of a typical star. Hence, stellar collisions are exceedingly rare within any one galaxy. Even when two giant galaxies collide, the star population density merely doubles, leaving ample space for the stars to meander without sustaining much damage. The stellar and interstellar contents of each galaxy are undoubtedly rearranged a little by the gravitational tides induced by collisions, but there are no spectacular explosions. The two galaxies just sort of glide through one another without causing much commotion.

Completing our inventory of the arrangement of matter in space, we arrive at the next obvious question: Are there even greater coagulations of matter in the Universe, or do galaxy clusters top the cosmic hierarchy? At present, astrophysicists are uncertain. There is suggestive evidence for clusterings of galaxy clusters, thereby shaping titanic galaxy superclusters, but this evidence is part conjecture, and hence subject to considerable debate. If correct, though, these presumptions mean that our Local Group along with several other galaxy clusters comprise a galaxy supercluster centered near the large Virgo Cluster. Galaxy superclusters, if they exist, are incomprehensibly gargantuan — on the order of several hundred million light-years in diameter, or more than a thousand times as large as the size of the Milky Way.

Please don't get swamped. Let's recapitulate for a moment. We live on planet Earth. Our planet orbits the Sun. The Sun is in turn just one of hundreds of billions of stars out in the suburbs of our Milky Way Galaxy. Our Galaxy is furthermore only one of many residents of the Local Group, which is in turn just an undistinguished galaxy cluster near the periphery of what might be an even larger galaxy supercluster.

There appears to be nothing special about our Earth, our Sun, our Galaxy, our Local Group. Evidently, mediocrity reigns throughout.

Such is our niche in the Universe.

∞

Astronomers have charted normal galaxies out to a distance of about four billion light-years. Many galaxylike objects are also known to exist beyond this galaxy horizon, but their fuzziness makes it tough to associate them with any of the normal galaxy classifications. More important, the basic character of the most distant objects seems to differ from that of those nearby. By and large, objects more than four billion light-years away are more "active," to a certain extent more violent. Overall, the emitting power of these active galaxies is much greater than that of the nearby spiral and elliptical galaxies. Furthermore, the active galaxies emit copious amounts of different kinds of radiation.

The adjective "normal," used to describe the ellipticals, spirals, and irregulars, is meant to convey that those galaxies simply emit the accumulated radiation of large numbers of stars. Most of their emitted energy is of the visible type, supplemented by lesser amounts of radio, infrared, ultraviolet, or X-ray radiation. This is not true of the active galaxies. Often, they emit enormous quantities of radio or infrared energy. Some are in fact completely invisible to us, undetectable with even the world's largest optical telescopes. Their presence is detected and studied by radio and infrared telescopes capable of capturing invisible radiation from space.

The emission of radiation from the active galaxies, then, is completely inconsistent with an accumulated emission of radiation by myriads of individual stars. To be blunt about it, it remains an open question as to whether active galaxies really have any stars.

The abnormal power and strange character of the predominantly distant astronomical objects suggest that the earlier epochs of the Universe were most violent. Remember, looking out in space is looking back in time. Since physical conditions were undoubtedly different in earlier epochs of the Universe, it shouldn't be surprising that distant astronomical objects that formed many billions of years ago differ from the nearby, recent ones. What is enigmatic — in fact, downright astounding — is the enormous amount of energy radiated from some of the most powerful objects. Their total release of energy often stretches astrophysical theory to its limits.

An average star, such as our Sun, emits at any one moment the equivalent of about a billion one-megaton nuclear bombs. Our Galaxy is, not surprisingly, a hundred billion times as powerful because, after all, it contains roughly a hundred billion stars. The active galaxies are generally a thousand times again as energetic.

Can you imagine the equivalent of a thousand normal galaxies all packed into the space usually occupied by one? This is the crux of the problem encountered while trying to explain the monstrously active galaxies. At one time, it was fashionable to suggest that these objects were the sites of spectacular galaxy collisions. However, calculations suggest that even such collisions could not produce energy in the amount required.

The fact that active galaxies often emit more radio and infrared than optical radiation suggests that these objects are fundamentally different from normal galaxies. Perhaps we shouldn't even be calling them galaxies.

The gross emission features of active galaxies can probably be understood by invoking a distinctly nonstellar phenomenon.

Called the synchrotron process, after the man-made accelerator machines used to study subatomic particles, this mechanism describes the emission of radiation when charged elementary particles interact with magnetic force fields. No stars are involved, nor is any heat per se. The radiation simply arises from magnetized regions of space.

Magnetism presumably pervades all things, not only Earth, Sun, and Solar System, but also entire galaxies. Although the magnetic force field in a typically diffuse galaxy is some thousands of times weaker than Earth's, magnetism can still play a significant role, especially when its effects mount across an entire galaxy.

Laboratory experiments have shown that when charged particles, particularly electrons, are injected into a magnetic force field, they spiral around much like the needle of a compass thrown through the air. Magnetism slows the particles, causing some of their kinetic energy to be transformed into radiant energy. The amount of radiation emitted from a single encounter of an electron and magnetism is not terribly large in the laboratory. But in the case of a great galaxylike object, the radiation can accumulate into vast quantities because of the huge number of electron encounters. Furthermore, the emitted radiation is often of the radio variety.

The general process aside, the details of the emission mechanism within active galaxies remain problematic, even assuming repeated injections of fast and numerous electrons. Still, the synchrotron mechanism gives us a glimpse of the type of events responsible for the emission of such vast amounts of radio power. All in all, observations of active galaxies imply a series of explosions that repeatedly accelerate electrons to speeds close to that of light itself. Then, in some cases, large clumps of plasma move outward, forming the distorted halos and extended blobs characterizing many of the active galaxies.

The upshot is that the theory of synchrotron radiation seems to be able to justify the observed emissions from typical active galaxies, provided they do not emit for much longer than a million years. However, the most violent of the active galaxies remain tough to explain, the variability of their radiation is nearly impossible to understand, the origin of their repeated explosions is currently unknown, and the reason for the injection of high-velocity electrons is downright mystifying.

Peculiar though they may be, the active galaxies are not the most energetic objects in the Universe. There is an additional, extraordinarily luminous class of active astronomical objects — objects so hard to fathom that they threaten to render inadequate the laws of physics as presently known. These are the innocuous-looking, though very distant and extraordinarily powerful, quasi-stellar sources — quasars for short.

Not content just to rival the energy emission difficulties of active galaxies, quasars actually extend those difficulties. The radio and optical radiation observed from quasars often displays variations from week to week, sometimes from day to day. The implication is straightforward: Galaxies could never synchronize their front-to-back emission to produce such rapid and coherent time variations. Expressed another way, no object can flicker in less time than it takes radiation to cross it. Thus, day-to-day variations imply that quasars cannot be much larger than a light-day, hardly more than the diameter of our Solar System. The enormous power of the quasars, usually a hundred to a million times that of our Milky Way Galaxy, must arise from a region tiny by cosmic standards.

Quasar emission mechanisms — whatever they may be — must then operate, by cosmic standards, within an almost unbelievably small realm of space. Can you imagine a hundred or more normal galaxies all packed into a region not much larger than the Solar System? That's an indication of the anomalous state of affairs required to appreciate the herculean quasars, unquestionably the most mysterious objects in all the Universe.

∞

Regardless of how they emit radiation from the depths of space, galaxies contribute little to the architectural design of the Universe. Galaxies are essentially "along for the ride," much like humans, who contribute little to the overall architecture of Earth as a planet. On the other hand, galaxies can be used to probe the framework of the Universe, in the same way that people can probe the structure of Earth. Galaxies are like billiard balls that can be used to determine the shape of a table top, or golf balls that can be used to survey the surface of a putting green. Cosmologists utilize the radiation and motions of distant galaxies to unravel the fabric of the Universe. Indeed, studies of already-formed normal and active galaxies are important if we are to appreciate the full realm of the cosmos, as has been already discussed in the Prologue.

Equally important are studies of the origin of galaxies. How did such enormous material coagulations arise from an early Universe comprising a mixture of only hot matter and intense radiation? Do galaxies form by engorging already-made stars, or do stars gestate in already-made galaxies? Which came first, stars or galaxies? Furthermore, how do galaxies evolve, once formed?

Fortunately, we can address these and other questions pertaining to the Matter Era with more assurance than we have about the rather uncertain events of the Radiation Era. Even here, though, there remain substantial puzzles about the details of the galaxy formation process. Scientists can address the problem and can identify the principal difficulties, but they cannot yet solve it completely.

Lack of a good observational understanding of the galaxies themselves creates the basic enigma. Galaxies can be classified according to their gross morphology and their total energy budgets. But we have as yet no explanation for the observed properties of all the galaxies in terms of, for example, the simple gas laws that describe our rather detailed understanding of stars. Not surprisingly, it's hard to fathom how the galaxies got there, when we don't quite know what they are.

Nonetheless, all galaxies have two common denominators. Together, these factors may help us understand the events that produced these most magnificent of all objects in the Universe.

First, no galaxies appear to be forming at the present time.

rthermore, none seem to have formed within the past ten billion
ars or so. Since all normal galaxies contain some old stars, and
ice all active galaxies are far away in space (and thus in time),
conclude that all observable galaxies have existed for a long
ne. Whatever the formation mechanism was, it was surely wide-
read in the early parts of the Matter Era. But if the galaxies
rmed so prolifically in the early Universe, then why aren't they
rming now?

A second common denominator derives from the observation
at most galaxies house comparable amounts of matter. The ca-
icity of virtually every individual galaxy thus far measured ranges
tween a billion and a trillion stars. Normal galaxies appear to
ive this many stars, while, as far as can be determined, active
ilaxies have the equivalent of this much matter. There are no
nown galaxies that are much smaller, and none that are much
rger. They all seem to have about a hundred billion stars, or
ieir equivalent, just like our own Milky Way Galaxy. Why should
ature's galactic coagulations have such a narrow range of sizes?
Vhat precludes the construction of galaxies containing, for in-
ance, a thousand trillion stars?

To address the questions of galaxy formation, imagine a giant
loud of hydrogen and helium atoms embedded in a weakening
ea of radiation, some tens of millions of years after the bang. This
iant cloud should not be thought of as filling the entire Universe,
nly a small sector of it. Physical conditions were changing rapidly
t the time. The Universal temperature and density had dropped
onsiderably with the onset of the Matter Era. Radiation was no
onger sufficiently intense to pyrolize atomic matter. In fact, fully
ormed hydrogen and helium atoms were becoming numerous
nough to exert a collective influence of their own. Electromag-
ietic and nuclear forces bound elementary particles within atoms,
vhile gravity in turn bound the atoms within the giant cloud. All
:he known forces that now direct the evolution of matter were
even then operating well enough to grant the cloud some integrity
of its own. Vast parcels of matter were becoming distinguishable

65

from other segments, a state of affairs strongly contrasting with t
stupendous and chaotic violence of the earlier Radiation Era.

Despite this stability, the initially homogeneous cloud wou
have surely experienced occasional fluctuations — small local
regularities in the gas density that came and went at random. N
cloud, whether a terrestrial fluffy cloud in Earth's atmosphere,
tenuous interstellar cloud in our Milky Way Galaxy, or tl
primordial cloud discussed here as part of the early expandi
Universe, can remain completely homogeneous indefinitely. Ea
of the cloud's atoms is sure to have some motion, largely becau
of heat. Eventually, one atom somewhere in the cloud will ac
dentally move closer to another, making that part of the clou
just a little denser than the rest. The atoms may then separat
dispersing this density fluctuation, or they may act together
attract a third atom to enhance it. In this way, small pockets of g
can arise anywhere in a cloud simply by virtue of random atom
motions. Each pocket is a temporary condensation in an otherwi
highly rarefied medium. Density fluctuations of this sort can als
cause fragmentation of a cloud into smaller assemblages. Th
whole process is not unlike the billowing clouds of a terrestri
thunderstorm, collecting and then dispersing their moisture.

Provided some density enhancements further develop by attrac
ing more and more atoms, they could conceivably grow int
groups of matter having the size of galaxies. Theoretical calcula
tions support the idea that such accidental gas fluctuations coul
have been the forerunners — protogalaxies — of present-day gal
axies. But — and this is an important but — these same calcula
tions suggest that, at this rate, the galaxies would only just b
forming at the present time. Yet, as we've seen, astrophysicist
have no evidence whatever that galaxies are now forming, hav
found no peculiar regions caught in the act midway between full
fledged galaxies and intergalactic nothingness.

A very long time — nearly twenty billion years — is needed fo
enough randomly moving atoms to coalesce into a large pocke
of gas that can be rightfully called a protogalaxy. The time re
quired is not surprising in view of the absolutely gargantuan quan

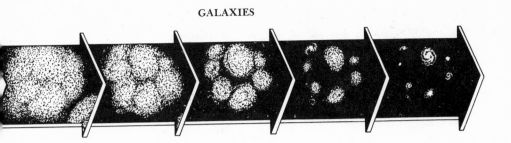

tity of atoms in a galaxy — namely, about a hundred billion stars per galaxy, times roughly a thousand billion billion billion billion billion billion atoms per star, or nearly a million billion billion billion billion billion billion billion atoms in a typical galaxy. That's 10^{68} atoms in scientific notation — clearly an awful lot of atoms to collect regardless of the notation used. Consequently, it takes a while for nature to do at random.

So, despite the fact that random enhancements in an otherwise homogeneous gas could have eventually produced galaxies, it's unlikely that the galaxies we now see originated strictly in this way. Still, the idea of naturally arising gas inhomogeneities remains a powerful concept, for it's a reasonably well understood process not requiring any unknown forces or unique conditions.

The problem of galaxy formation is currently a tough one for astrophysicists. Its solution has exasperated many brilliant minds. The origin of galaxies appeals to theorists having fertile imaginations, and especially to those willing to make unorthodox assumptions. It's one of the most tricky areas of astronomy to appreciate, for few hard facts are known about galaxies, and fewer still about the physical mechanisms that formed galaxies long ago.

One hard fact that is clearly known, however, is that galaxies do exist. And they exist in great abundance. Somehow they got there. Let's then consider in greater detail some of the specific galaxy formation mechanisms recently proposed by theoreticians.

Some researchers have embraced a radical viewpoint and have argued that perhaps the stars formed first, after which they clustered to fabricate the galaxies. In their opinion, galaxies did not originate by means of the traditional hypothesis of random in-homogeneities in the gas-radiation mix of the early cosmic fire-ball. Instead, they propose that the galaxies originated much later by conglomerating stars, a model generally known as gravitational clustering. This unorthodox view, then, suggests that galaxies could indeed be forming at the present time, but partly from old stars that formed billions of years ago.

According to this theory, first the planets and then the stars formed from primitive gas fluctuations present in the early years of the Matter Era. These were gaseous irregularities within an otherwise rapidly cooling and completely homogeneous fireball. Stars having formed, gravity bundled them together, fashioning star clusters. These clusters in turn migrated to form the galaxies, after which the galaxy clusters and the galaxy superclusters formed in succession.

The attractiveness of this hypothesis lies in its hierarchical pack-age: All big objects are successively constructed from small ones. This sort of material buildup is just the reverse of the material breakdown expected if the Universe were to contract in the future. Apart from this pleasing theoretical symmetry, though, practical problems plague the concept.

One difficulty with the gravitational-clustering hypothesis is that — we must repeat — there is no evidence that galaxies are form-ing at present. If they were taking shape only now, we should be able to observe objects having a morphology somewhere between well-defined galaxies and sheer empty space. We know of no such amorphous, undefined objects. Furthermore, the regions out be-yond the galaxy clusters — the intergalactic medium — do not appear to contain much matter. Whenever and however the galaxies did form, they apparently did so very efficiently, sweeping up almost all the material available, and leaving nothing behind for further conglomeration. In addition, strong theoretical argu-ments suggest that stars ought to be forming now within the galaxies. These arguments have been handsomely verified within the past decade by splendid observations of widespread locations in our own Milky Way, where stars are now known to be originating slowly but surely from the galactic hodgepodge of gas and dust. Yet another problem is that it's nearly impossible to imagine the formation of planets prior to the formation of stars. This is especially true if, as seems reasonable, the heavy elements comprising planets are synthesized in the cores of stars. The original stars, almost certainly, preceded the original planets.

The crux of the issue here is that reasonably good observational evidence can virtually prove that stars are currently forming and that galaxies are not. Most contemporary arguments and all modern data point straight toward the notion of an early formation of galaxies, followed by a later formation of stars and planets within those galaxies.

Not willing to accept mildly radical hypotheses like gravitational clustering of stars or extreme hypotheses suggesting that galaxies originate from the expansion of small dense objects, most astrophysicists today embrace some version of the random gas fluctuation concept discussed earlier. Remember, theory suggests that these fluctuations did indeed arise, but that their growth would have been too slow to fabricate galaxies before now. If some way could be found to accelerate their growth, the problem might be solvable. One such way is to assume that the Universe was quite turbulent long ago.

Turbulence was probably an important factor in the early Universe. By turbulence, we mean the inevitable "confusion" or disordered motion of matter (the gas) within a rapidly moving medium (the Universe). All the atoms within the vast primordial clouds were set into motion, not only from the expulsion of the bang, but also from the heat of the fireball. The gas then had some "directed" kinetic energy — outward, from the ordered expansion of the Universe. It also had some "undirected" kinetic energy — random, from the disarrayed aftermath of the blazing inferno. Intact pockets of gas undoubtedly surged this way and that, whirling round and round in addition to the individual agitations of each of the atoms. In particular, turbulence helped to drive swirling eddies at locations where density fluctuations had already become established in the early Universe.

Turbulent eddies of this sort can be visualized by watching water swirl down the drain of your bathtub. Even better examples can be noted by moving your hand gently through water, or a teaspoon through coffee; swirling eddies naturally materialize in the wake of this turbulence.

There are probably no better examples of turbulence than in the fluffy clouds of Earth's atmosphere. Especially vivid in photographs of the tops of the clouds, taken with Earth-orbiting satellites, kilometer-sized whirling eddies can be noted as enhancements in the atmospheric gas. These eddies are known to become more pronounced whenever air currents are particularly turbulent. Should they grow, in this case by accumulating additional amounts of moisture, the eddies may well form hurricanes hundreds of kilometers across. Some of them do, though fortunately for us infrequently.

Here is a case, then, where studies of a terrestrial phenomenon — Earth's weather — may help us understand one of the most vexing extraterrestrial problems. Planetary hurricanes roughly mimic the overall morphology, the pancake shape, the differential rotation, and the concentration of energy within spiral galaxies. These several resemblances suggest that we may be able to learn something about galaxy formation via the study of hurricane formation. In particular, since most meteorologists agree that some sort of turbulent "priming" is required to initiate a hurricane, the early stages of such storms could conceivably be used by astronomers to extract some clues about the elusive density fluctuations that gave rise to protogalaxies in the early Universe.

Despite the constant cooling caused by the expansion of the Universe, each localized eddy within a large gas cloud must begin to heat. It can't avoid it. Eddies are sites, not only of turbulence, but also of increasing heat within a steadily cooling cloud. The heat results from friction caused by frequent collision among the increasingly dense collections of atoms within each eddy. It's not unlike the heat derived by rubbing our hands together on a cold winter day.

Eventually, individual eddies must rid themselves of some of this newly acquired energy, much as the Sun or any other heated object needs to unload energy, lest it blow up. The eddies in the protogalactic cloud do it by radiating their heat. In this way, a large cloud containing lots of eddies can cool even faster than would a normally homogeneous cloud in the expanding Universe. As it cools, the entire cloud contracts a little, thereby increasing the density and hence the heat within the eddies. Individual eddies and the whole cloud simultaneously radiate this newly gained energy into space, thereby allowing further contraction of both the parent cloud and the individual eddies. On and on, this cycle of contracting, heating, radiating, cooling, and contracting proceeds. The cycle may operate at different speeds for each of the eddies, particularly since some eddies will be more successful than others at sweeping up additional gas from the parent cloud.

It's easy to conceptualize a cluster of galaxies forming in this way, with each eddy becoming a member galaxy within that cluster. Alternatively, perhaps only one or a few galaxies form within each of the vast primordial clouds of the early Universe, after which gravity gradually sweeps the galaxies into the incomprehensibly large galaxy clusters now seen scattered throughout the abyss.

As nice as this galaxy-formation scenario seems, it runs into some problems once mathematics is applied to it. Detailed calculations can be made for the amount of time required for these individual eddies of gas to contract into objects resembling galaxies. Here the principal difficulty arises. The time required for coagulation and contraction of the gas in a turbulent eddy is longer than the typical time for the random dissipation of that eddy. In other words, the eddies are predicted to break up long before they have a real chance to coagulate tightly. Turbulent eddies do enhance the random gas fluctuations, but they don't last long enough to form galaxies.

Any kind of eddy, then, comes and goes in an iffy sort of way. Eddies appear, disappear, and reappear at different parts of either a terrestrial atmospheric cloud of moist air or an extraterrestrial galactic cloud of primordial gas. Occasionally, a terrestrial eddy does indeed grow to form a flourishing hurricane, or an extraterrestrial eddy presumably a full-fledged galaxy. But the expected rarity of their rapid growth implies that turbulent eddies cannot be the sole solution to the problem of formation of the galaxies or of galaxylike objects.

As mentioned earlier, most contemporary researchers avoid radical theories of galaxy formation. They prefer to work with the basic notion of random gas fluctuations. But some mechanism must be found to speed the growth of such fluctuations in the cooling and thinning primordial Universe. The current problem then centers on other ways that might have enhanced the growth of gas fluctuations. Modern theorists have tried to solve just this problem for years, though they have not really succeeded. Galaxy formation is indeed a tough nut to crack.

The general scenario now favored by the astronomical community postulates that the early Universe was not perfectly homogeneous. Instead, the Universe is theorized to have been peppered, even at the start of the Matter Era, with substantial density fluctuations. These already-formed pockets of gas could then have swelled during the galactic epoch to fabricate at least the basic outlines of today's galaxies. The accepted mechanism of galaxy formation is considered to be the familiar gravitationally induced cycle of contracting, heating, radiating, cooling, and eventual flattening into disk-shaped objects. Significantly, the presence of gas density fluctuations toward the end of the Radiation Era accelerates the process and shortens the formation time. In all truthfulness, however, an inhomogeneous Radiation Era leaves a bad taste in the mouths of some researchers.

Much of the fascination experienced by workers studying galaxy formation derives from the fact that many of the theories cannot be proved incorrect. Many ideas remain possible, there being no experimental data to the contrary. People familiar with the sophisticated mathematics of notoriously tough subjects such as fluid mechanics, turbulent physics, and magnetohydrodynamics justify their interests by tinkering with the problem of galaxy formation. Yet, despite considerable efforts in the past decade or two to unlock the secrets of galaxy formation, the specifics of a plausible process have thus far eluded discovery.

∞

Whatever the galaxy formation process might have been, it or some subsequent evolution led to the myriad galaxies now observed. There are loose and tight spiral galaxies, with mixtures of old and new stars. There are large and small elliptical galaxies, containing only old stars. And there are irregular and explosively active galaxies, let alone the baffling quasars, which may not house any stars at all.

With such a zoo of galaxylike objects in mind, it's natural to wonder if there is any overall pattern or evolutionary scheme interrelating the various types of galaxies. The answer is, none that we know of at present. There is no known physical operation underlying all the galaxies.

Some researchers long ago suggested an evolutionary progression of galaxies starting with the near-spherical ellipticals, which became flattened ellipticals, eventually changing into closed spirals, followed by open spirals, and finally culminating in irregular galaxies. The central idea here contends that galaxies originate with a more or less spherical shape and, as they grow older, their rotation tends to flatten them, first producing some ellipticity, and gradually some spiral arms, prior to their breaking up as aged irregular galaxies. This type of evolutionary hypothesis requires that all elliptical galaxies be young and all irregular galaxies old. But this is not the case at all. Observationally, elliptical galaxies are not young. They're old, nearly depleted in interstellar gas and dust, and displaying no evidence of active star formation.

On the other hand, we might argue that since ellipticals are clearly old galaxies, then perhaps the evolutionary scheme progresses in the opposite sense. Maybe irregulars are young and, having formed first, gradually evolve into ellipticals. It's easy to imagine loose spiral galaxies wrapping up into tighter spirals and eventually into elliptical galaxies. But difficulties abound here, too. Apart from the obvious problem of understanding how the beautiful spiral galaxies might have been fabricated from the contorted irregular galaxies, there remains the need to reconcile this theory with the abundance of old stars in the irregular and loose-spiral galaxies. Simply put: If irregular galaxies and loose spirals are the starting point in a scheme of galaxy evolution, then all of them should be young. But they're not. Virtually all irregulars and spirals contain a mix of old and new stars. The existence of old stars is just not consistent with the nature of a youthful galaxy.

The upshot is that normal galaxies probably do not evolve from one type to another. Spirals really do not seem to be ellipticals with arms. Nor do ellipticals appear to be spirals without arms. There are no known parent-child relationships among the normal galaxies. Their dispositions apparently result exclusively from the physical conditions extant in the gas clouds from which they originated more than ten billion years ago.

An evolutionary link between normal galaxies and active galaxies seems more credible, though still debatable. A time sequence proceeding from quasars to active galaxies to normal galaxies implies a continuous range of cosmic objects. Adjacent objects along this sequence are almost indistinguishable from one another. For example, weak quasars have a few things in common with some very active galaxies, while the feeblest active galaxies often resemble the most explosive normal galaxies. Perhaps all galaxy-sized objects started out more than ten billion years ago as quasars, after which they gradually lost their emissive powers, becoming in turn active galaxies and eventually normal galaxies.

This idea stipulates that the observed quasars are actually ancestors of all the other galaxies. Far too distant for us to see any stars, the quasars are detectable only because of their tremendously energetic central regions. Likewise, because of their great distance, we perceive them as they once were, in their blazing youth. As the

central activity decays with time, quasars assume forms closer to those of more familiar and nearby galaxies. Should this view be proved, then even our Milky Way Galaxy was once a brilliant quasar.

Although attractive, this quasar → active galaxy → normal galaxy evolutionary hypothesis has its drawbacks. Some researchers argue that there is no evolutionary link at all. They suggest that the powerful quasars are merely extreme manifestations of the explosive phenomena seen in virtually all galaxies. After all, even the center of our own Milky Way is known to be expelling matter and radiation. The same can be said for active galaxies and quasars, though on a larger scale. Perhaps all these objects are part of the same family without there being any evolutionary sequence linking its members, just as there is no evolutionary transformation among different races within the human species. Each galaxy type or human race is distinctly different. One race of humans does not evolve into any other, and similarly one type of galaxy may not necessarily evolve into any other. Instead, they all may be essentially ordinary galaxies that formed long ago, some having been endowed with especially explosive central regions. Those able to exercise their explosiveness more than others for some as yet unknown reason are called quasars, while those hardly able to explode at all are called normal galaxies. Why the quasars explode most frequently and violently is not known. The answer presumably lies buried within the relatively uncharted centers of galaxies.

It seems that future research directed toward the centers of galaxies will provide the most help in unraveling the secrets of these gargantuan cosmic objects. Whether all galaxies actually change from one type to another, or some simply undergo repeated outbursts while remaining the same type of unevolved object, is thus far unsolved.

∞

Our knowledge of the galaxies, especially their origin and evolution, is clearly inadequate. How each of them got there, endowed with its peculiar shape and prodigious energy, remains largely a mystery. The problem is intensified by the fact that astronomers cannot find any galaxy in the act of formation. Furthermore, even if galaxies do evolve, their changes are sure to be so slow, compared to the length of time our technological civilization has been studying them, as to make them appear immutable.

The origin and evolution of galaxies tender more problems than the formation of stars, which we can observe; than the evolution of stars, which we can identify; than the origin of life, with which we can experiment in our laboratories; than the evolution of life, which we can study in action; than the origins of intelligence, culture, and technology, which we can probe using fossils unearthed from deep layers of historical rubble. Practically everything else discussed in this book is on firmer ground than the origin and evolution of galaxies. Indeed, the subject of galaxy formation is the biggest missing link in the scenario of cosmic evolution.

Galaxies, though, are very important. Aside from the creation of atoms, the formation of galaxies was the first great accomplishment of the Matter Era. Until we gain a great deal more knowledge about them, our understanding of cosmic evolution will remain incomplete and unsatisfactory.

STARS

Forges for Elements

billions of years ago

STARS ARE BALLS OF GAS, tenuous and hot on the outside, dense and hotter on the inside. Except for their shape, they do not resemble hard, rocky planets in any way whatever. Normal stars are enormously larger and tremendously hotter than planets, and experience changes in a completely different manner. They have no real surface, let alone any hard, solid matter like that on Earth. Stars are simply composed of loose gas held intact by the relentless pull of gravity.

This same gravity forces the gas to take on the simplest possible geometrical configuration — a sphere. Wherever gravity dominates, it compels matter to adopt a round shape. All the known stars, planets, and moons are spheres.

Gravity is not the only force operating in stars. Otherwise, the inward pull of this all-pervasive force would shrink stars to such a small size that they could not radiate heat and light. Competing against gravity in a star is the pressure of the heated gas. Pushing outward, this pressure tries to disperse the star into space. The result is an equilibrium, or stable condition: gravity in, pressure out. That's the simple prescription for a star — any star.

The star we know best is called the Sun.

Stars are interesting astronomical objects for many reasons, though two stand out. The first is easy: Stars are the furnaces where heavy elements are forged. Without the heavies, nothing around us — not the ground, not the air, no part of the Earth — could exist.

Second, stars play an essential role in the heating and lighting of nearby planets. In the case of our own Solar System, for example, the energy of our Sun is a critically important factor that not only led to the development of life on Earth, but that also yields the heat and light necessary for the continued maintenance of that life. The Sun, more than anything else, makes Earth a reasonably comfortable abode. Without a nearby star, life as we know it could not exist.

Stars, then, are enormously significant in the evolution of both matter and life. Clearly, it's of some import to understand how, when, and where stars originate and evolve.

Ultimately, stars may also help us comprehend one of the foremost questions of all, namely, the enigmatic events of that singular point of superhot, superdense matter out of which the Universe originated. Burned-out corpses of massive stars, popularly known as black holes, are thought to mimic, more than any known astronomical object, the bizarre conditions prevalent at the very start of the Universe. Also of importance, then, is an understanding of how, when, and where stars die.

Astrophysicists know a great deal about how stars are born, exist, and die, how they pass through phases of youth, maturity, and aging. Although they appear immutable in the nighttime sky, stars actually change their appearance throughout their life cycles. Some stars are old, some young. Others are long gone, having literally run out of gas and died eons ago. Still others have yet to be fashioned from the interstellar hodgepodge of gas and dust. We don't notice all of this change because stellar life cycles are astronomically long compared to human life spans.

∞

The origin of stars is of fundamental scientific significance.
ere are few things more basic than the birth of a star. Do astro-
ysicists really know all the steps in the formation of a typical
r? Can we describe the specific evolutionary paths of the inter-
lar inhomogeneities that eventually produce stars? Questions
e these are now being answered. The solutions are not yet com-
tely clear, but remarkable progress has been made in the past
ade. As it stands now, knowledge of star formation is a com-
ation of theoretical insight and proven fact.

The stellar epoch offers a better description of matter than does
earlier galactic epoch. Gravitational instability arguments, in-
ed with partial success for galaxies, can be used more profitably
understand the formation of stars within those galaxies, regard-
of how the galaxies originated. As was the case in the early
iverse, random fluctuations can occur at various parts of a large
cloud within any already-formed galaxy. Although these fluc-
tions proved insufficient by themselves to cluster huge amounts
matter into galaxies, theoretical calculations suggest that the
cess should work well enough to coagulate smaller chunks of
tter into stars. Swirling eddies in the medium between the
s are cooler and denser than those of the primordial fireball,
l hence are better suited to collect the amount of matter re-
ired to mold individual stars or groups of stars. Generally, stars
thought to form when sufficient matter accumulates in a galac-
pocket, after which it contracts, heats, and eventually ignites
internal nuclear fire.

Astrophysicists have constructed intricate models of the stages
ough which gas clouds evolve to become genuine stars. These
lar models are essentially "numerical experiments" performed
large, high-speed computers. For the stellar epoch, the compu-
ional factors include gravity, heat, rotation, magnetism, chemi-
reactions, elemental abundances, and a few other physical
ditions typifying a contracting interstellar cloud. These factors
like the ingredients of an elaborate recipe. In this case, the
ipe is mathematical, teeming with complex equations. And as

79

is true for any new recipe, the types of ingredients are known, l
the amounts of each are often uncertain.

Large computers built during the 1970s enable theorists to
a trial-and-error procedure for this multifaceted problem — to
sure that the recipe works. Though computers do nothing m
than crunch numbers, they can do this basic task more quic
than humans, adjusting the many ingredients while maintain
consistency between the theoretical predictions and the obser
tions of the myriad stars within our Milky Way Galaxy.

The accuracy of these stellar models is presently unknown,
it's tricky to take that third step of the scientific method and
them experimentally. To be sure, no one has ever seen an int
stellar cloud or a genuine star parade through its evolution.
paces. The lifetime of a human, or even of our entire civilizati
thus far, is very much shorter than the time necessary for a clo
to contract and form a star. Since about thirty million years
about a million human generations) are required to concoct a s
such as our Sun, no person can realistically hope to observe a
astronomical object proceed through its full pageant of star bir

Our models are not without observational support, howev
Telescopic examination of various gas clouds at different stages
their evolutionary trek helps refine our overall comprehension
star formation. Newly invented technology enables experimen
ists to probe interstellar clouds and nascent stars for hints ab
their embryonic development. By studying these various int
stellar clouds, often at unrelated locations in our Milky W
Galaxy, astrophysicists have been able to piece together an ob
vational understanding of many of the important stages of
stellar evolution.

Current efforts of astrophysicists resemble those of archeolog
and anthropologists who unearth artifacts or bones at numer
unrelated localities strewn across our planet's surface. Not hav
had the opportunity of living at the time of our ancestors, th
Earth scientists ponder the myriad remains, trying to fath
how all of them might be pieced together into an overall pict

of human evolution. Similarly, space scientists observe various unrelated objects of our Galaxy, seeking to understand how each object fits into the overall scheme of stellar evolution. The terrestrial bones and extraterrestrial stars are much like the pieces of a puzzle. The picture will become clear only when each segment is found, identified, and oriented properly relative to all the other pieces.

∞

Imagine a large region of interstellar matter someplace in our Galaxy. By definition, *inter*stellar matter is that which exists out beyond each of the stars. Most people think nothing resides there, for, sure enough, a clear night shows only blackness among all the minute points of light we know as stars. But the darkness of outer space only affirms the limits of our vision. Not much matter resides in any one place, to be sure, but it does indeed exist. Interstellar space is so vast that even small amounts of matter scattered here and there accumulate to play a significant role. It's not unlike the prospect of becoming a multimillionaire by collecting a mere penny from every person in the United States. Even minute quantities can accumulate into exceedingly large amounts, given enough space and time.

The interstellar medium, then, is the rarefied galactic region from which all stars arise at birth, and into which many stars explode at death.

Matter in interstellar space is usually a mixture of gas and dust. Most of the gas is in the form of atoms, though there are a few clusters of atoms — namely, molecules — scattered about. Interstellar dust particles are not terribly unlike the fine chalk dust that settles on blackboard ledges, or domestic dust that lurks under beds and in closets; tiny particles in a terrestrial fog might be better examples.

If the gas and dust of interstellar space had remained evenly dispersed forever, neither stars nor planets — and certainly not life

— would ever have formed. The sky would be absolutely dark, and no one would exist to know it. Fortunately, the interstellar medium is not immutable. Like everything else, it changes its disposition.

Theory suggests that matter contained within the dark regions of space develops fluctuations and thus fragments into large clumps typically spanning tens or even hundreds of light-years. Because these dark regions are just that — dark — they have always been difficult or impossible for astronomers to discern. Quite frankly, there is nothing to see in a dark region — which, by the way, partly explains why scientists have been, until recently, virtually ignorant since the birth of astronomy about star formation.

Dark and dusty regions of interstellar space are inaccessible to study by optical astronomers. Even stars behind these regions are invisible because dust diverts their radiation from reaching Earth, hopelessly scattering it like automobile headlights in a fog. That doesn't mean that the murky interstellar recesses are totally impenetrable, however. The marvels of modern technology permit invisible regions to be sampled for the infrared and radio radiation they emit. These alternative types of radiation have longer wavelengths than ordinary light does, enabling them to penetrate the debris of interstellar space. With the same techniques that soldiers use to locate the enemy at night by using infrared sensors, and by the same means that radio receivers operate properly in the foggiest of weather, radio and infrared researchers can detect invisible radiation from sheer darkness in interstellar space.

Analysis of the radiation emitted by interstellar matter has now experimentally confirmed our theoretical prediction that much of the near-void among the stars of any galaxy is clumped into large gassy clouds. Their overall morphology tends to resemble the irregular, fluffy clouds of Earth's atmosphere, but there the resemblance ends. Interstellar clouds are billions of times as large as the entire Earth; they also materialize and disperse billions of times as slowly as terrestrial clouds.

Radio and infrared observations have established that interstellar clouds are cold and tenuous, often containing no more than a hundred atoms per cubic centimeter. This density is extremely low, in fact far lower than that of the best vacuums attainable in physics laboratories around the world; for comparison, the normal density of air on Earth is more than a billion billion atoms per cubic centimeter. Their typical temperature, some —250 degrees Celsius, is also extremely low, for the lowest possible temperature (at which all atomic motion ceases) is —273 degrees Celsius.

Now imagine a small portion of an interstellar cloud, for instance a parcel of gas and dust much less than a light-year in diameter. Because of the cloud's tenuosness, such a parcel does not house many atoms. Unless the cloud is as cold as physically possible, each atom will have some random motion.

While moving around, each atom is influenced somewhat by the gravitational force exerted by all the other neighboring atoms. This force is not very large, owing to the small mass of each atom. In fact, even if a few atoms were to coalesce accidentally for a moment, their combined gravitational pull would be insufficient to bind them permanently into a distinct condensation. The accidental cluster would disperse as quickly as it formed.

Suppose we widen our sights to include more than just a few atoms. Instead, consider fifty, a hundred, or even a thousand atoms. Would a group of that many atoms exert a combined gravitational force strong enough to prohibit them from dispersing as in the previous example? Just how many atoms are needed for gravity to bind them into a tight-knit coagulation?

Answers to these questions cannot be found from a simple study of gravity alone. The correct solution depends, not only on gravity, but also on several other physical conditions such as heat, rotation, magnetism, and turbulence. These additional agents tend to influence the evolution of an interstellar cloud, for, although they should not be regarded as antigravity, they do compete with gravity.

Analyses of radiation emitted by galactic regions prove that interstellar clouds have some heat, though not much. Most of their warmth derives from inevitable collisions among the atoms. More frequent atomic collisions mean greater friction and thus more heat, just as rapidly rubbing our hands together generates more warmth than doing so sluggishly. Heat gives a cloud of gas some buoyancy that tends to compete with gravity. Heat is, in fact, the principal reason that the Sun doesn't collapse; the outward pressure of its heated gas counteracts the inward pull of gravity. The amount of heat contained within interstellar clouds is, of course, small by solar or even terrestrial standards. Consequently, thermal effects, which compete strongly with gravity once stars are formed, do not really play a large role until after interstellar clouds begin contracting and thus generating greater amounts of heat.

Rotation — that is, spin — can also compete with the inward pull of gravity. A contracting cloud having even a small spin tends to develop a bulge around its midsection. This bulge is a sure indication that some of the matter is trying to defy gravity and thus disperse. As the cloud contracts in size on its way to becoming a star, its spin necessarily increases, just as a figure skater rotates more rapidly with her arms retracted. Any rapidly rotating object exerts an outward force; the faster the spin, the greater the force. Anyone can feel this outward force while bearing the brunt of many circular rides at amusement parks. In the case of an interstellar gas cloud, atoms near the periphery are particularly vulnerable to outward escape should the inward pull of gravity prove insufficient to retain them. If the rotation of a contracting gas cloud were to increase to the point at which gravity could no longer keep it intact, then the cloud would simply disband, restoring its atoms to the more tenuous interstellar medium. Mud flung from a rapidly rotating bicycle wheel is a good example of this. The only way an interstellar cloud can preserve itself against the threat of outward dissipation via rotation is to gather more and more atoms, thereby increasing the collective strength of gravity. The upshot is this: Rapidly rotating interstellar clouds need more mass to guarantee continued contraction toward starlike objects than do clouds having no rotation at all.

Magnetism, turbulence, and several other physical effects can also hinder the contraction of a gas cloud. Observations made during the past decade show that real interstellar clouds are not very hot, spin only slowly, and are only slightly magnetized and turbulent. But theory suggests that even minute quantities of any of these agents can compete effectively with gravity. Surprisingly small amounts of each can unite to alter considerably the evolution of the typical gas cloud.

So it's not a simple case of gravity sweeping up matter to form a star. There are many additional factors serving to complicate the problem, making the process tricky to understand in detail.

We now return to our original question: How many (hydrogen and helium) atoms need to be accumulated for the collective pull of gravity to prohibit a pocket of gas, once formed, from dispersing into the surrounding interstellar space? The answer, even for a cool cloud having no rotation or magnetism, is a very large number. In fact, nearly a thousand billion billion billion billion billion (i.e., 10^{57}) atoms are required for gravity to bind a gaseous condensation. There's no question about the truly huge magnitude of this number. It's much larger than the number of grains of sand under all the oceans of the world, even larger than the million billion billion billion billion billion (i.e., 10^{51}) elementary particles comprising all the atomic nuclei throughout the entire Earth. It's large compared to anything with which we're familiar because there's simply nothing on Earth comparable to a star.

This number, 10^{57} atoms, is equivalent to just about the mass of our Sun. This is no coincidence. Our Sun is a very ordinary star, implying that most stars form from interstellar fragments having approximately this number of atoms. In all, stars can originate from slightly larger or smaller condensations, for the range of known stars varies from about one-tenth to one hundred times the mass of our Sun, in astronomical terms a small variation.

Spanning ten to a hundred light-years, typical interstellar clouds usually harbor thousands of times as much matter as do normal stars. Observations prove this. If a cloud is to become the birthplace of stars, it cannot remain as an intact homogeneous blob. Interstellar clouds must gradually break up into subcondensations, often less than a light-year across. Theory suggests that fragmentation into subunits occurs naturally, because gravitational instabilities at various parts of an interstellar cloud force the development of inhomogeneities in the gas. A typical cloud is then expected to break up into tens, even hundreds, of fragments or clumps, each imitating the shrinking behavior of the cloud as a whole, albeit contracting even faster than the parent cloud.

There is no evidence for stars born in isolation, one star from one cloud. Interstellar clouds are thought to begin their long evolutionary trek either to form numerous stars, each much larger than our Sun, or whole clusters of stars, each comparable to or smaller than our Sun. Most stars, perhaps all, originate as members of star clusters. Those now appearing alone and isolated in space, such as our Sun, probably wandered away from the rest of their litter, though only after all were fully formed.

Once a fragment takes on its own identity within an interstellar cloud, it then passes through a series of inevitable stages. It first begins to contract as gravity affects the ever-accumulating group of atoms. It literally shrinks under the pressure of its own weight. As the protostar becomes more compact, the atoms collide more frequently, causing the gas fragment to warm.

By the time a typical self-heating fragment has shrunk to about a tenth of a light-year across, its temperature has risen to nearly zero degrees Celsius. That's still colder than our twenty-degree-Celsius room-temperature standard, but it's a lot warmer than the original interstellar cloud prior to this clumpy stage. Also, the size of the fragment at this stage has diminished considerably, but it's still some hundred times the size of our Solar System. Such a gas clump must change still further, reorganizing itself into a smaller, denser, hotter object, before it can be rightfully called a star.

Our description is more than a theoretical scenario. It has now been clearly, though not visually, substantiated, using equipment only recently developed. In the 1970s, radio and infrared observations began to produce direct evidence that interstellar clouds are fragmenting into smaller clusters of gas. Pockets of slightly hotter and slightly denser matter within the otherwise tenuous, cold, and enormous interstellar clouds are now known to be the rule rather than the exception.

Fragmentation might be expected to continue indefinitely, producing ever-decreasing clumps that couldn't possibly form stars. Fortunately, the process halts before it's too late. Increasing gas density stops the process of fragmentation from reducing all parts of the cloud without limit into ever-smaller subunits. As individual fragments compress their gas, they eventually become compact enough to prohibit radiation from easily escaping. With the cloud's natural vent partially blocked, the trapped radiation causes the temperature to rise, pressure to increase, and fragmentation to cease.

As each gas condensation continues to evolve, various theoretical models predict much the same story: The fragment's size diminishes, its density grows, and its temperature rises at both the core and the periphery. Several thousand years after it first began contracting, a typical fragment's dimensions will have become comparable to those of our full Solar System, a size still ten thousand times as large as our Sun. Internal temperatures at this stage will have reached many thousands of degrees Celsius, temperatures greater than those within the hottest steel furnaces fashioned by our civilization here on Earth.

Some hundred thousand years later, the full expanse of a fragment could fit within Earth's orbit around the Sun. And its temperature has steadily mounted to nearly a hundred thousand degrees Celsius. That's clearly a lot of heat. Elementary particles, now ripped from disintegrated atoms, are really whizzing around inside. Despite this veritable inferno, though, those particles are still too sluggish to overcome their natural electromagnetic repulsion in order to enter into the realm of the nuclear force. In other words, the matter is still far from the ten million degrees Celsius required to initiate nuclear burning that will one day transform this gaseous heap into a genuine star. Nonetheless, the hot, dense object at this stage resembles a star closely enough to merit the special name protostar. Such an embryonic object is perched at the dawn of star birth.

Theoretical models aside, is there any observational evidence that hot, dense fragments have Solar System dimensions? Indeed there is. Within the past decade, radio telescopes have captured radiation emitted by small clumps at or near the core of many cloud fragments. The diameter of the clumps is hardly more than a thousandth of a light-year, or just about equivalent to the size of our Solar System. Their total gas densities reach nearly a billion particles per cubic centimeter. And their temperatures have been measured by infrared techniques at many hundreds of degrees Celsius. Most astrophysicists agree that these dense, warm blobs are genuine protostars, poised on the verge of stardom.

Protostellar objects emit radiation, much of which comes from small molecules formed when a few atoms link together. The radiation is especially intriguing because of its terrifically high intensity. The first observations in the 1960s of the radio radiation from one such blob were so mysterious that puzzled researchers began calling the molecular emitter "mysterium." It was later identified properly as the hydroxyl (hydrogen plus oxygen) molecule, and these enormously powerful signals are now known to be enhanced or amplified by a very special "maser" process.

The word *laser* has become a common everyday term. It's actually an acronym for *l*ight *a*mplification by *s*timulated *e*mission of *r*adiation. Lasers are experimental devices that emit a concentrated stream of light radiation in a very narrow beam. Only within the past couple of decades has our civilization become smart enough to build such tools, relying as they do on both advanced technology and a good understanding of atomic and molecular physics. Lasers operate by exciting atoms or molecules in a gas, and then stimulating them to emit radiation simultaneously. In this way, a tremendous burst of radiation can result, much more powerfully than from ordinary light bulbs.

Masers are similar to lasers, except that they produce *m*icrowave (low-frequency radio) radiation rather than optical radiation. We can build them in our terrestrial laboratories, though masers are very delicate machines, requiring special conditions and much patience to operate. When working properly, they are the best amplifiers known, much more effective than ordinary transistors.

Interestingly enough, certain regions of interstellar space are naturally suited to produce amplified microwave radiation. Protostellar blobs apparently enjoy the special conditions required, first, to excite some molecules and, second, to stimulate them to emit intensely. The blobs' warm temperatures and moderate densities seem ideal for this unique emission mechanism. Accordingly, the intense maser radiation observed from certain molecules can be analyzed for clues about protostellar regions. Such studies comprise one of the most exciting areas of contemporary astrophysics.

Theory suggests that protostars are still a little unstable. The inward pull of gravity does not quite balance the outward push of gas pressure. Fortunately, the temperature is still too low to establish that "gravity-in, pressure-out" equilibrium that guarantees stability. We say "fortunately," because if the heated gas were able to counteract gravity before reaching the point of nuclear burning, then there would be no stars. The nighttime sky would be abundant in dim protostars, though completely lacking in genuine stars. And it's likely that neither we nor any other intelligent life forms could exist to appreciate them.

Computer models predict that as protostars continue to contract, the gas has no alternative but to follow the dictates of gravity. The result is renewed heating. But even with a core temperature of a million degrees Celsius, there is still not enough heat to initiate nuclear burning. Only when the temperature deep down in the core reaches ten million degrees Celsius do the nuclear reactions commence. Atomic nuclei then have enough thermal energy to overwhelm their own repulsion by means of the very same violent process described earlier for the transformation of hydrogen into helium during the particle epoch. A star has finally formed. Its principal function thereafter is to consume hydrogen, thereby producing helium and especially energy.

Hearts of stars, then, are sites where atomic nuclei viciously collide, penetrating the realm of the nuclear force, thus releasing copious amounts of energy. Every second, our Sun releases more energy than humans have generated in all of history. The energy from the nuclear inferno moves up through the interior of the star and is radiated from the surface in the guise of starlight and other types of stellar radiation. The starry points of light seen in the nighttime sky owe their existence to the nuclear fires churning deep in the cores of each one of them. Ponder all that astronomical activity while looking upward some clear, quiet evening.

The remarkable change from interstellar cloud to contracting fragment to protostellar blob to nascent star takes about ten million years. Obviously a long time by human standards, this is still less than one-tenth of one percent of a star's complete lifetime. The entire process amounts to a steady metamorphosis, an evolution of sorts, a gradual transformation of a cold, tenuous, flimsy pocket of gas into a hot, dense, and round star. The prime instigator in all this evolutionary change is gravity.

∞

Once heat and gravity are balanced, a star like our Sun is stable. It will steadily produce energy for some ten billion years. A combination of theory and experiment suggests that the Sun has already done so for about five billion years. So it can be thought of as a middle-aged star, a celestial body expected to burn steadily for another five billion years into the future.

Stars smaller than our Sun take more time to form from interstellar matter. They also last longer. For example, stars having one-tenth the mass of the Sun require hundreds of millions of years for birth, and are expected to endure for as long as a trillion years.

Since this is much longer than the age of the Universe, all small stars that have ever formed must still be merrily burning hydrogen into helium, producing a constant flux of energy for any attendant planets.

Stars larger than our Sun tend to form faster from interstellar clouds, some in as short a time as a million years. In contrast to small stars, big ones seem to do everything at a quickened pace. They burn their hydrogen fuel more rapidly as well; their greater mass gravitationally compacts large stars more than others, causing the matter within to collide more frequently and violently. This in turn hastens the nuclear reactions. Despite their magnitude, the biggest stars endure for far less than the ten-billion-year lifetime of our Sun. The most massive ones, for example, are nearly a hundred times the mass of the Sun, but last for only about ten million years. They expend their stability with a great flurry, a mere wink of an eye in the normal scale of cosmic lifetimes. Alas, a quickened pace is not always a desirable one. Just as it's unhealthy for humans to rush through life, the largest stars hardly seem to settle down at all. In the end, while small stars shrivel up and fade away, stars more massive than our Sun perish by catastrophic explosion. Apparently some clichés have Universal applicability: The bigger they are, the harder they fall.

∞

Hardly anything notable befalls a star during its normal lifetime. Provided the nuclear fires continue to balance the relentless onslaught of gravity, nothing spectacular happens to the star as a whole. Predictably, its core transforms hydrogen into helium, its surface erupts in flares and spots, and its atmosphere releases vast amounts of radiation capable of influencing the history of any attendant planets. But, by and large, stars do not experience any sudden changes while in equilibrium. They simply fuse hydrogen into helium during this, the longest phase in the history of any star, a phase lasting about ninety-nine percent of their lifetime.

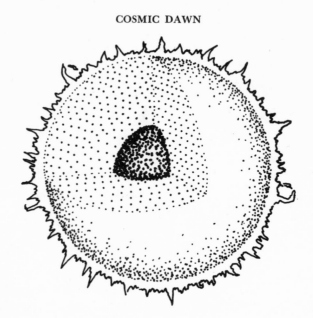

Stars, in equilibrium, would continue to produce energy indefinitely, pending some drastic change. Eventually, something drastic does occur in all stars: they begin running out of fuel.

Computer modeling is again our foremost guide to the specific changes experienced by any star near death. Identifying numerous physical and chemical factors, and adjusting their magnitudes again and again, theoreticians have built models to account for the wide variety of stars seen in the real Universe. Let's first detail the death plunge of a star like our Sun, after which we can broaden the discussion to include all stars, large and small. Keep in mind, though, that all these fatal events occur within the last one percent of a star's lifetime.

Theory suggests that, as the Sun ages, its hydrogen steadily becomes depleted, at least in a small, central core about a hundredth the star's full size. After nearly ten billion years of slow and steady burning, little hydrogen will remain within the zone of ten million degrees Celsius. It's probably a bit like an automobile that can cruise along a highway at a constant velocity of fifty-five miles per hour for many hours without a care in the world, only to have

the engine cough as the gas gauge approaches empty. Unlike automobiles, though, stars are not easy to refuel.

Widespread exhaustion of hydrogen in the stellar core causes the
nuclear fires there to cease. Hydrogen combustion continues unabated in intermediate layers, above the core though well below
the surface. But the core itself normally provides the bulk of the
support in a star, establishing its foundation and guaranteeing its
equilibrium. The lack of core burning assures instability because,
although the outward gas pressure weakens in the cooling core,
we can be sure that the inward pull of gravity does not. Gravity
never lets up. Once the outward push against gravity is relaxed —
even a little — changes in the star become inevitable.

Generation of more heat could bring the aged star back into
equilibrium. If, for example, the helium at the core could begin
to fuse into some heavier element such as carbon, then all would
be well once again, for energy would be created as a by-product to
help reestablish the outward gas pressure. But the helium there
cannot burn — not yet, anyway. Despite a phenomenal temperature of more than ten million degrees Celsius, the core is just too
cold for helium to fuse into any heavier elements.

Recall that a temperature of at least ten million degrees Celsius
is needed to initiate the simpler hydrogen → helium fusion cycle.
That's what it takes for two colliding hydrogen nuclei to get up
enough steam to overwhelm the repulsive electromagnetic force
between two like charges. Otherwise, the nuclei cannot penetrate
the realm of the nuclear binding force, and the fusion process
simply doesn't work. Well, with helium, even ten million degrees
Celsius is insufficient for fusion. Each helium nucleus has a net
positive charge twice that of the hydrogen nucleus, making the
repulsive electromagnetic force much greater. To ensure successful fusion by means of a violent collision between helium nuclei,
exceedingly high temperatures are required. How high? About a
hundred million degrees Celsius.

Lacking that degree of heat, the star's core of helium ash does
not remain idle for long. Once its hydrogen fuel becomes substantially depleted, the helium core begins contracting. It has to;

there's not enough pressure to hold back gravity. This very shrink-age allows the gas density to increase, thereby engendering more heat as gas particle collisions become ever more frequent.

The increasingly hot core continues to heat the intermediate layers of this stellar furnace. It's a little like turning up the stove from warm to hot. Higher temperatures — at this stage, well over ten million degrees Celsius — cause hydrogen nuclei in the inter-mediate layers to fuse even more rapidly than in the core before.

The aged star is really in a predicament now. The core is un-balanced and shrinking, on its way toward generating enough heat for helium fusion. The intermediate and outermost layers are also unbalanced, fusing hydrogen into helium at a faster-than-normal rate. The gas pressure exerted by this enhanced hydrogen burning grows greater, forcing the intermediate layers and especially the outermost layers to expand. Not even gravity can stop them. Even while the core is shrinking, the overlying layers are expanding! Clearly, the former equilibrium state is completely ruined.

The two observable aspects of such a perverse star are interest-ing. To an astronomer far away, the star would seem gigantic, nearly a hundred times as large as normal. Captured radiation would also suggest that the star's surface was a little cooler than normal. (This is not to say that the act of either ballooning or chilling of an aged star could be directly observed. Theory sug-gests that the transition from a normal star to an elderly giant takes much longer — about a hundred million years — than a human lifetime.)

The second change — surface cooling — is a direct result of the first change — increased size. As the star grows larger, its sum total of heat is spread throughout a considerably larger stellar volume. Hence, visible radiation emitted from such a cooling, but still-hot, surface shifts in color. Like a white-hot piece of metal that turns red while cooling, the whole star also displays a reddish tint. Over the course of time, long by human though short by stellar stan-dards, a star of normal size and yellow color slowly transforms into one of giant size and red color. The bright normal star has evolved into a dim red giant.

So, once the Sun exhausts the hydrogen fuel supply at its core, instability is sure to set in. The core will shrink. The overlying layers will expand. In short, the Sun is destined to become a swollen sphere a hundred times its normal size, in fact large enough to engulf many of the planets, including Mercury and Venus, and perhaps Earth and Mars as well.

There's no need for humans to panic, not yet at any rate. Provided the theory of stellar evolution is reasonably correct as described here, we can be sure that our Sun will not balloon into this red-giant stage for another five billion years.

Red-giant stars are not figments of some theoretician's imagination. They really exist, scattered here and there about the sky. Even the naked eye can perceive the most famous of all red giants — a star called Betelgeuse, a swollen, elderly member of the constellation Orion.

Should the inherent imbalance of a red-giant star be maintained unabated, the core would eventually implode, while the rest of the star receded in slow motion. Various forces and pressures at work inside such a decrepit star would literally pull it apart. Fortunately for the stellar veteran, this torturous shrinkage-expansion is not expected to continue indefinitely. Within a hundred million years after the star first begins to panic for lack of hydrogen fuel, something else happens — helium begins to burn. Though this seems like a whole new lease on life, it amounts to only a brief reprieve.

Deep down inside a red-giant star, the density increases as the interior pressure builds. Once the matter in the star's core has become a thousand times as dense as at the center of a normal star, collisions among the gas particles will be violent and frequent enough to generate sufficient heat, via friction, to reach the hundred-million-degree temperature required for helium fusion. Helium nuclei henceforth collide, suddenly transforming into carbon nuclei, and igniting the central fires once again. Thereafter for a period of a few hours, the helium burns ferociously, like an uncontrolled bomb.

Despite its brevity, this nuclear reaction releases an enormous flood of new energy. The energy is potent enough to etherealize the core matter somewhat, thereby lowering the density and relieving some of the pent-up tenseness among the charged nuclei. This small expansive adjustment of the core halts the gravitational collapse of the star, returning it to equilibrium — an equilibrium once again between the inward pull of gravity and the outward push of gas pressure.

Once the helium → carbon fusion reactions have commenced, thus stabilizing the core, the hydrogen → helium fusion reactions churning in the layers above subside a bit. We might want to think of the star as having expanded its outer layers too rapidly, overshooting the distance at which the star achieves a comfortable gravity-heat equilibrium throughout. The entire star is then able to shrink a little, losing its swollen appearance. Like all the other evolutionary changes in the early or late phases of a star, this slight size adjustment is made quickly — in about a hundred thousand years.

Though the time scales for marked stellar change are considered rapid throughout all phases of a star's emergence from dust as well as its thrust toward doom, we stress that they are still long compared to the duration of a human lifetime. Observers have little hope of watching an individual star move through the evolutionary paces characterizing the stellar epoch. Instead, we must rely on theoretical computations to provide a good approximation of the varied stages in the birth and death of a star.

This reliance on computer modeling is exactly what makes the results of an important experiment so disturbing. The one experiment that bears directly on the physical processes inside stars does not jibe well with the predictions for a star like our Sun. In particular, scientists have been unable to detect the expected number of neutrino elementary particles in the solar radiation reaching Earth. Derived from an Italian word meaning "the little neutral one," neutrinos are known from experiments on Earth to be virtually massless and chargeless, and to travel at the velocity of light. Interacting with almost nothing, neutrinos are endowed with an ability to pass freely and unhindered through a thickness of several million light-years of lead! Hence, they should be able to escape without hesitation from the solar core, where they are created as a by-product of nuclear reactions. Ordinary radiation bounces around in the solar interior for millions of years before being emitted to space, but neutrinos are predicted to arrive at Earth a mere eight minutes after they're made. They thus constitute the only direct test of the nuclear mechanisms thought responsible for powering the Sun.

Neutrinos must nonchalantly penetrate Earth at all times; though we are neither aware of nor harmed by them, they apparently pepper our bodies constantly. However, one of the rare materials with which neutrinos do interact is a chemical with the tongue-twisting label, tetrachloroethylene. As complex as it sounds, it's a safe fluid commonly used in the dry-cleaning industry. To count the flux of solar neutrinos arriving at Earth, a "neutrino telescope" has been constructed at the bottom of a gold mine by filling a large tank with a hundred thousand gallons of this stuff. Depth is necessary to assure the experimenters that other elementary particles from nonsolar events do not interfere. Although the experimental apparatus seems to be working properly, the rate of neutrino detection is considerably less than theory predicts.

Astrophysicists are currently wrestling with the results, attempting to resolve what looms as a serious threat to our knowledge of stellar fusion. Both theorists and experimentalists, however, are reluctant to attribute the underabundance of solar neutrinos to any conceptual errors in the theory of stellar evolution. Some suspect the experimental apparatus, others the computer models. Still others argue that we don't yet know enough about the odd neutrino particle itself. All researchers nonetheless concur — or

97

at least hope — that the correct interpretation of the neutrino experiment will not demolish our understanding of solar fusion.

There remain limitations to every aspect of cosmic evolution. Here in the stellar epoch, as elsewhere, researchers seem to know the broad outlines of many things, but the fine details are often yet to be understood.

The nuclear reactions in a star's helium core churn on, but not for long. Whatever helium exists in the core is rapidly consumed. The helium → carbon fusion cycle, like the hydrogen → helium cycle before it, proceeds at a rate proportional to the temperature; the greater the core temperature, the faster the reaction progresses. Under these very high temperatures, then, the helium fuel in the stellar core simply doesn't last long — no longer than a few million years.

Buildup of carbon ash in the inner core causes physical phenomena similar to those in the earlier helium core. Helium first becomes depleted at the very center. Fusion then ceases there, after which the carbon core shrinks and heats a little, causing the hydrogen and helium burning cycles to speed up in the intermediate and outermost layers of the star. These layers ultimately expand, much as they did earlier, making the star once again a red giant.

Provided the core temperature becomes high enough for the fusion of two carbon nuclei, or even a mixture of carbon and helium nuclei, still heavier products can be synthesized. Newly generated energy again supports the star, returning it to its accustomed equilibrium between gravity and heat.

This contracting-heating-fusing-cooling cycle is generally the way in which many of the heavy elements are fashioned within the last gasps of stellar cores. All elements heavier than helium are created within the final one percent of a star's lifetime.

∞

How do stars die? Again, we must rely partly on computer modeling, and partly on what is observable in the sky. The problem, quite frankly, is that no one has witnessed a star die since the invention of the telescope nearly four centuries ago. Guided by theoretical predictions of how stars behave near death, astronomers search the Universe, seeking evidence of objects resembling the predicted hulk.

All theoretical models suggest that the final stages of stellar evolution depend critically on the mass of the star. As a rule of thumb, we can say that low-mass stars die gently, whereas high-mass stars die catastrophically. Our Sun and all smaller stars are members of the low-mass category, while stars much larger than our Sun are grouped in the high-mass category.

The demise of our Sun is destined to be rather straightforward and unspectacular. The Sun's core will become extraordinarily compact and hot. A single cubic centimeter of stellar core material will eventually weigh a ton. That's a thousand kilograms of matter compressed into a volume the size of a thimble. Yet, even at such high densities, collisions among nuclei are insufficiently frequent and violent to raise the temperature to the phenomenally high six hundred million degrees required to fuse carbon into any of the heavier elements. There's simply not enough matter in the over-lying layers of the smaller stars to bear down any harder. Conse-quently, oxygen, iron, gold, uranium, and many other elements are not synthesized in low-mass stars.

Small stars like our Sun manage to work themselves into quite a pickle in their old age. Their carbon core is, for all intents and purposes, dead. Helium just above the core of carbon ash con-tinues to transform into more carbon, while hydrogen in the inter-mediate layers converts into more helium. This onslaught of heating slowly pushes away the outermost layers to ever greater distances. The theoretically expected upshot is an object of weird posture having two distinct parts. Called a planetary nebula, it is predicted to have a halo of warm, tenuous matter completely veil-ing a hot, dense core.

Nebula is Latin for "mist" or "cloud" of great extension and extreme tenuity. The adjective "planetary" is plainly misleading, for these astronomical objects are not related to planets in any way.

The designation dates back to the eighteenth century, when optical astronomers could barely distinguish among the myriad faint, fuzzy patches of light in the nighttime sky. Some researchers mistook all of them for planets. But subsequent observations clearly demonstrated that the nebula's fuzziness results from a shell of warm gas surrounding a small central star. Modern telescopes fully resolve these planetary nebulae, enabling astronomers to recognize their true nature.

Weird or not, nearly a thousand examples of planetary nebulae have been discovered in our Galaxy alone. Direct observations confirm the theoretical predictions that the shell consists of an envelope in the act of being gently expelled from the core of an aged red-giant star.

Further discussion of the evolution of the expanding envelope is not very interesting. It simply continues to spread out as time passes, becoming more diffuse and cool, and eventually merging imperceptibly with the interstellar medium. This is one way, then, that the interstellar medium becomes enriched with additional helium atoms and possibly some carbon atoms as well.

Continued evolution of a core, or stellar remnant, at the center of a planetary nebula, is not much more exciting. Formerly concealed by the atmosphere of the red-giant star, cores make their first appearance once the flimsy envelope has receded. These cores are small, hot objects, highly abundant with carbon, but no longer experiencing nuclear burning. They shine only by stored energy, though their small size guarantees a white-hot appearance. Not much bigger than planet Earth, they are called white-dwarf stars.

Analysis of the radiation emitted by white-dwarf stars shows their properties to agree pretty well with the computer model predictions. Many are found at the very center of planetary nebulae. Several hundred additional ones have been discovered naked in our Galaxy, their envelopes blown to invisibility long ago.

So astronomers are able to identify red-giant stars, planetary nebulae, and white-dwarf stars in the nearby cosmos. At different stages in their old age, each of these objects seems to match the

overall disposition predicted by the theoretical calculations for elderly low-mass stars. Once again, though, we should not expect to witness the act of envelope expulsion during the course of a human lifetime. It takes several tens of thousands of years for a red giant's atmosphere to recede sufficiently for a white-dwarf star to appear.

Nothing exciting happens to dwarf stars after this. For all practical purposes, these "stars" are dead. They continue to cool, becoming dimmer with time, slowly transforming from white dwarfs to yellow dwarfs and then red dwarfs. The final state is that of a black dwarf — a cold, dense, burned-out ember in space. Such stellar corpses have reached the graveyard of stars.

No one knows how many black dwarfs really exist in the Galaxy. That's not surprising since they're after all unlit. Even if these dark clinkers could somehow be detected, we would probably find that there aren't many of them. The total lifetime of a low-mass star is long — comparable to or longer than the age of the

Galaxy. Our Milky Way has not been around long enough for many low-mass stars to have fulfilled the whole cycle from birth to death. Perhaps none has.

A different fate awaits objects having more than several times the mass of our Sun. By and large, they evolve much like their low-mass counterparts up through the red-giant stage, with only one difference. All the evolutionary changes occur more rapidly for high-mass stars because their larger mass enables them to generate more heat. And, more than anything else, heat speeds all evolutionary events.

At the red-giant stage, the core of a high-mass star is able to attain the six hundred million degrees Celsius required to fuse carbon into even heavier elements. Large mass is the key here. Massive stars generate a stronger gravitational force than solar-type stars, and the added gravity can crush matter in the core to a high enough density to ensure frequent and violent collisions among the gas particles.

Theoretical models call for a highly evolved star of large mass to have several layers where various nuclei burn. The interior of the star resembles an onion. At the relatively cool periphery just below the surface, hydrogen fuses into helium. In the intermediate layers, helium and carbon fuse into heavier nuclei. Just above the core, there reside magnesium, silicon, sulfur, and numerous other heavy nuclei. The core itself is full of iron nuclei, rather complex pieces of matter each containing several dozen protons and neutrons, and situated midway between the lightest and heaviest of all known nuclei. Each of the fusion cycles, during which nuclei for new elements are created, is induced by periods of stellar instability. The core cools, contracts, heats some more, fuses into heavy nuclei, becomes depleted of fuel, cools again, contracts again, and so on. At each stellar-burning stage, energy is released as a by-product of the fusion process, effectively supporting the star against gravity.

With iron in the core, complications develop for this sick and dying star. Nuclear interactions involving iron do not produce energy; iron nuclei are so compact that energy cannot be extracted. To a certain extent, iron nuclei play the role of fire extinguisher, suddenly damping the stellar inferno, at least at the core. With the appearance of substantial quantities of iron, the fires terminate for the last time.

Potential for disaster now clearly exists. No longer is this very massive star upheld by nuclear burning at the core. The star is suddenly without foundation; equilibrium has completely vanished. Even though the temperature in the iron core has by this point reached several billion degrees Celsius, the enormous inward gravitational pull of matter ensures catastrophe in the star's near future. Unless nuclear fires continue unabated, trouble is a certainty for any star.

Once gravity overwhelms the pressure of the hot gas, the star implodes, falling in on itself. The implosion doesn't take long, perhaps only hours after cessation of core kindling. Internal temperatures and densities then rise phenomenally, causing the star to rebound instantaneously, detonating parts of its core while jettisoning all the surrounding layers. Much of its mass — including a variety of heavy elements cooked within — is expelled into neighboring regions of space. The expulsion is much, much more violent than that in a planetary nebula; this star has exploded catastrophically. All stars much larger than our Sun are slated to perish in this way. Such a spectacular death rattle is known as a supernova.

Nova is Latin for "new." Astronomers now recognize that supernovae are not really new stars at all. Their sudden brightening, briefly transferring them from invisibility to prominence, only made them appear so to the ancients.

A supernova outburst is one of the most tumultuous events in any galaxy. The exploded stellar debris is hot and altogether can radiate a flash equal to nearly a billion times the brightness of our Sun. This amounts to a single star suddenly blazing to a billion times the Sun's luminosity within a few hours after the outburst.

Eventually, as in any explosion, the surge subsides, but not before the galactic neighborhood has been irradiated with plenty of potent energy and heavy elements.

Astrophysicists are unsure of the details at the exact moment of explosion of a supernova because a nearby star has not erupted in this way since Galileo Galilei first used a lens as a telescope early in the seventeenth century. Nor are the theoretical models entirely clear on the intricacies of the explosion.

These models suggest that, while heavy elements such as carbon, nitrogen, oxygen, sodium, magnesium, silicon, and iron are produced in stellar interiors, the explosion itself is responsible for elements heavier than iron. At the moment of explosion and for about fifteen minutes thereafter, some intermediate-weight nuclei are fiercely jammed together, thus creating the heaviest of all nuclei. Many of the rare elements are synthesized at this time, including nickel, silver, gold, uranium, and plutonium. Matter most valued by Earth societies therefore originated in the very last gasps of stars once big and bright, though now dead and disintegrated. Ironically, the heaviest of all elements are fashioned only after the large stars have perished.

The sprinkled debris of erstwhile stars then mingles with fresh interstellar hydrogen and helium produced during the earliest epoch of the Universe. This mixture of all elements can then undergo contraction, heating, and nuclear burning, thus fabricating second, third, and nth-generation stars in a seemingly endless cycle of death and rebirth — a kind of cosmic reincarnation. Our own Sun is at least a second-generation star, for it already contains heavy elements. Since these heavies could not have been created in a low-mass, relatively cool star like the Sun, they must be the products of formerly massive stars that exploded long ago.

How do we know that stars really do cook heavy elements in this way? How can we be sure that this theory of stellar nucleosynthesis is correct? We're assured by one piece of circumstantial evidence, and one item of direct evidence. First, the rate at which various nuclei are captured and the rate at which they decay are known from laboratory studies performed during the 1960s and 1970s. When all these rates are incorporated into a computer program, which also takes account of the temperatures, densities, and compositions at many positions within a typical star, the relative amounts of each type of synthesized nucleus match fairly closely the relative abundances of the ninety-odd elements found in nature. This is especially true for the abundances of intermediate-weight elements up to and including iron. Thus, despite the fact that no one has ever directly observed atomic nuclei in the act of production, we can be reasonably sure that the theory of stellar nucleosynthesis makes sense in view of our knowledge of nuclear physics and stellar evolution.

Second, observation of one type of nucleus — a very heavy one named technetium — provides direct evidence that heavy-element formation really does occur in the cores of stars. This nucleus is known from laboratory measurements to have a radioactive half-life of about two hundred thousand years. This is a very short time astronomically speaking, and hence the reason why no one has ever found even traces of naturally occurring technetium on Earth; all of it decayed long ago. (It can be manufactured and studied, however, in nuclear laboratories.) On the other hand, observation of technetium in numerous red-giant stars implies that it must have been synthesized within the past several hundred thousand years.

Supernovae are not just idle predictions of theoreticians. There's plenty of evidence that cosmic explosions came to pass in our Galaxy long ago. One of the most heavily studied remnants of a supernova is the Crab Nebula, so named largely because its appearance resembles that type of marine animal. About five thousand light-years from Earth, its debris is strewn over a ten-light-year extent. Now greatly dimmed, the Crab Nebula can be seen only through a large telescope. But the measured motions of the expelled matter — traveling at some thousand kilometers per second

— imply a brilliant explosion whose radiation must have first arrived here some nine hundred years ago. Indeed, the original explosion was so spectacular that ancient Asian and Middle Eastern manuscripts claim its brightness rivaled the Moon's in the year 1054 A.D. The Crab could be seen in broad daylight for nearly a month.

Numerous other massive stars no doubt vomited in earlier times. Ancient records document the emergence of at least a half-dozen supernovae in our Galaxy within the past thousand years. The nighttime sky harbors additional evidence for many remnants of stars that must have blown up well before the advent of recorded history. Furthermore, astronomers patrolling the skies with telescopes occasionally notice a sudden brightening of a portion of some faraway galaxy, enabling them not only to verify that high-mass stars are common to all galaxies, but also to refine the predictions of the theoretical models. Nearly a hundred such supernovae have been observed in other galaxies during the twentieth century. Their brilliance, mimicking on a much larger scale the basic features of human-made thermonuclear bomb blasts, often momentarily rivals that of the entire distant galaxy in which they explode.

The last supernova observed in our Galaxy caused a worldwide sensation in Renaissance times. The sudden appearance and subsequent fading of a very bright stellar object in the year 1604 A.D. helped shatter the Aristotelian philosophy of an immutable Universe. Little did anyone then realize that this exploded star provided the mental seedlings for the eventual emergence of the scenario of cosmic evolution, wherein the concept of change is central.

The infrequence of supernovae is a bit disturbing. Knowing the rate at which evolutionary steps are theorized to occur and estimating the number of massive stars known in the Galaxy, we might expect a galactic supernova to pop off in an observable location (away from the dusty parts of the Milky Way) every hundred years or so. It's unlikely that any such garish explosions could have been missed since the last one a few centuries ago. Hence, the Milky

Way seems long overdue for a supernova. Unless massive stars explode much less frequently than suggested by the theory of stellar evolution, we should be treated to nature's most spectacular event any day now.

Supernovae may be more than splendid light shows. Should a massive star detonate in the galactic suburbs where our Sun resides, it could well inundate Earth with radiation harmful to life. Indeed, the topmost layers of massive stars are ripped off and sent flying into space in the form of extremely fast-moving elementary particles, colloquially named cosmic rays. An understanding of the physical properties of the nearby stars is thus of more than just passing interest. An ability to predict the manner in which the nearby stars will die is downright critical. Of particular concern is the possibility of one of our neighboring stars exploding as a supernova, although we probably couldn't do much about it even if one did.

Inventories of stars in our galactic neighborhood suggest that a supernova can be expected to happen within thirty light-years of our Sun once every half-billion years. Too close for comfort? Fortunately, none of the closest stars is massive enough to die catastrophically by exploding. Luckily for us, they all seem destined to perish, as will our Sun, via the more placid red giant–white dwarf route.

Almost certainly, a viewable massive star has *already* exploded, but the light from this stupendous event has yet to reach our planet. Should such a supernova suddenly appear in the sky, we can be sure that all the world's major astronomical instruments will immediately focus in the direction of this, the grandest of light shows. Many observatories have in fact established "supernova alert teams." The Harvard-Smithsonian Observatory, for example, has several astrophysicists ready to commandeer, within an hour's notice, all ground-based telescopes and orbiting spacecraft operated by this joint establishment. (A false alarm on a recent Labor Day even helped smooth out some communications hazards, should a supernova inconsiderately time its earthly arrival on a human holiday!) The prime objective will be a study of the early phases of a supernova outburst, especially the various types of emitted radiation stretching from relatively harmless radio waves to potentially lethal gamma rays.

∞

What remains in the aftermath of a supernova explosion? Is the entire star just blown to bits and ejected into the surrounding interstellar medium? Not really. Most theoretical models predict that some portion of the star survives. As with planetary nebulae that expel matter less violently, supernovae also leave a core of remnants. Material within this severely compressed core comprises one of the strangest states in all the Universe.

During the moment of implosion of a massive star, just prior to its explosion, all the electrons in the core violently smash into the protons. The electrons have been there all along, but the protons are freed when some of the heavy nuclei disintegrate under the phenomenal onslaught. The result is an elementary-particle reaction that proceeds throughout the core of the massive star, systematically converting within seconds all the electrons and protons into neutrons and neutrinos. The neutrinos rapidly leave the scene at the velocity of light; they are suspected by many theorists of playing a major role in the triggering of supernovae, for neutrinos must transport much of the energy of the collapsed core to the overlying layers of the star, deposit it there, and cause the rest of the star to discharge into space. Much heavier than the neutrinos, the material debris departs at speeds much less than the velocity of light. Only the core remains intact. It has become a ball of neutrons. Researchers colloquially call this core remnant a neutron star, but it's not really a "star" in the true sense of the word.

The theory of stellar evolution predicts that neutron stars are very small, though still massive. Composed purely and simply of neutron elementary particles touching one another in a jam-packed sphere, a neutron star is not much larger than a typical city. Each unexploded core remnant usually equals several times the mass of our Sun, making neutron stars extraordinarily com-

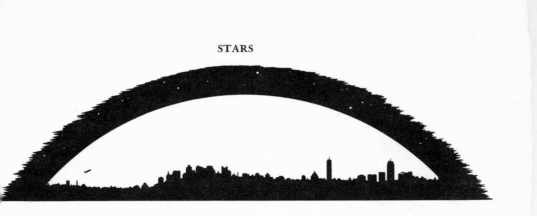

pact. Their average density is estimated to reach at least a trillion times that of Earth rocks. Not just a huge density, this is an incredible density, nearly a million times as dense as the already super-compact white-dwarf stars. In fact, the density of a normal atomic nucleus is not much greater; neutron stars are just about as compressed as the matter within the nuclei of normal atoms. Potentially significant, such extraordinary density was already encountered in the tale of cosmic evolution during the earliest particle epoch of the Universe. Detailed study of weird neutron stars might therefore enable scientists to understand better the physical conditions prevalent just after the start of the Universe.

Once stars explode as supernovae, all nuclear reactions cease. Theoretical calculations suggest that the remnant neutron stars are solid objects, more like planets than stars. Provided they've cooled sufficiently, we might imagine standing on one. It wouldn't be easy; a neutron star's gravity is unbelievably intense. A person weighing an Earthly one hundred fifty pounds (seventy kilograms), and standing on the surface of a neutron star, would weigh the Earth equivalent of about a million tons (a billion kilograms). Actually, standing wouldn't even be possible, for the severe pull of gravity would level a person to the thickness of a postage stamp. Gravity is so strong on a neutron star that, if shipped there, the population of the world would be crushed into a volume the size of a pea!

Can we be sure that objects as strange as neutron stars actually exist? The answer is again yes. In the past decade, radio and X-ray observers have made some remarkable discoveries, proving that neutron stars are for real. These researchers monitored several hundred pulsating stars, or pulsars for short, that emit short radiative pulses lasting for about a hundredth of a second apiece. Each pulse contains a burst of radiation, after which there is nothing. Then another pulse arrives. The time intervals between pulses are astonishingly uniform — so much so, in fact, that the repeated emissions could be used as a clock.

Many pulsars appear to be directly associated with supernova remnants. The most intensively studied pulsar resides close to the center of the Crab Nebula. By determining the velocity and direction of travel of the ejected matter observed for that supernova remnant, researchers have then been able to work backward, pinpointing the location in space at which the explosion presumably occurred. There the supernova core remnant is expected to be located. And that's precisely the region in the Crab Nebula from which the pulsating signals arise. Apparently, pulsars are the remains of the once-massive stars.

Astrophysicists reason that the only physical mechanism consistent with such precisely timed pulsations is a small, rotating source of radiation. Only rotation can cause the high degree of regularity of the observed pulses. And only a small object can account for the sharpness of each pulse; radiation emitted by an object larger than about ten kilometers in diameter would arrive at Earth at slightly different times, blurring the sharpness of the pulse. It's hardly surprising then that the best theoretical model of a pulsar envisions a small, compact, spinning neutron star that periodically flashes radiation toward Earth. The experimentalist's "pulsar" and the theoretician's "neutron star" are synonymous.

According to a leading theoretical model, a "hot spot" on the surface of a neutron star, or in the atmosphere above it, continuously emits radiation in a sort of narrow searchlight pattern. This "spot" could be a violent surface or atmospheric storm much like the less energetic flares on our Sun, or volcanoes on Earth. Although the spot sprays radiation steadily into space, the star's spin rate of many times per second guarantees that the emitted radiation in any direction behaves like a hail of discrete bursts. The radiation sweeps through space as if from a revolving lighthouse beacon. Arriving at Earth, perhaps thousands of years later, it's observed as a series of rapid pulses. The duration of each pulse carries information about the source of activity on the neutron star, while the period of the pulses reveals the star's rotation. The details of the theoretical model are sketchy and controversial, for researchers have hardly any information about the behavior of matter having a density as great as a million tons per cubic centimeter.

Neutron stars are indeed outlandish objects. Theory, though, predicts that they are more or less in equilibrium, just like most other stars. In the case of neutron stars, however, equilibrium does not mean a balance between the inward pull of gravitational force and the outward pressure of hot gas. Neutron stars have no hot gas. Instead, the outward force is provided by the crystalline nature of the tightly packed neutrons. Existing side by side, the neutrons form a hard ball of matter that not even gravity could compress further — with one notable exception.

Suggestions have been made that all galaxies harbor stellar core remnants with masses so large that the inward pull of gravity can in fact overwhelm even the seemingly incompressible sphere of pure neutrons. According to some theories, should enough matter be packed into an exceedingly small volume, then the collective pull of gravity could gradually crush any countervailing phenomenon. In this case, gravity is envisioned to be so powerful that it can compress a massive star into an object the size of a planet, a city, a pinhead, a microbe, even smaller! The gravitational pull in the vicinity of these objects is thought to be so great that light itself would return to them much as baseballs return to Earth upon being thrown into the air. Such freakish objects would be expected to emit no light, no radiation, no information whatever. Incommunicado, such a massive star would have effectively collapsed into a hole — a hole perhaps no larger than a few centimeters across, but a hole into which all nearby matter falls, trapped by gravity perhaps forever. Astrophysicists call these most bizarre end points of stellar evolution black holes.

∞

A black hole is a region containing a huge amount of mass spread throughout an extremely small volume. It's not an object per se so much as a hole, and one that's dark to boot. The hole's gravitational force field is fantastically large — great enough to warp space and time severely in its vicinity. A black hole has enormous gravity partly because of its huge mass. But that's only half of the law of gravity. The other half dictates that gravity is inversely proportional to the square of the distance. And because the distance term is squared, the gravitational force field can grow spectacularly if the distance between any two parts of an object is decreased, that is, by compressing the object further.

When the surviving mass of a supernova core remnant exceeds several solar masses, no known force can counter gravity. Not even neutrons, touching one upon another, can halt the pull of gravity within such a compact object. Theory suggests that the former star collapses indefinitely, crushing matter to the dimensions of a point. It catastrophically implodes without limit; apparently nothing can stop it.

Can anyone possibly appreciate such a seemingly ridiculous phenomenon? How can a star shrink to the size of an elementary particle, while presumably on its way to still smaller dimensions? Does this make sense? Well, this is indeed the prediction arising from detailed mathematical formulations. Without some agent to compete against gravity, massive core remnants are expected to be instantaneously squeezed into singular points of infinitely small volume.

Though we cannot here set forth the exceedingly complex mathematics necessary to understand the true nature of black holes, it's nonetheless possible to discuss a few qualitative aspects of these phenomenally dense and highly eccentric regions of space.

Consider first of all the concept of escape velocity. For any rela-

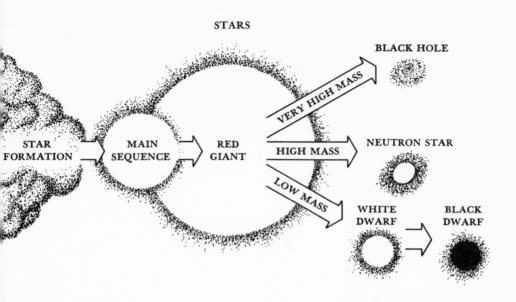

tively small piece of matter — molecule, baseball, rocket, whatever — the velocity needed to escape from a larger object is proportional to the square root of the larger object's mass divided by its radius. For example, on Earth, with a radius of about six thousand kilometers (or four thousand miles), the escape velocity equals about ten kilometers per second (or seven miles per second). To boost anything from the surface of our planet, a velocity greater than this must be achieved.

Imagine now a hypothetical experiment for which the principal apparatus is a large three-dimensional vise. Let the vise be large enough to hold the entire Earth. As awful as it sounds, imagine Earth to be squeezed on all sides. As our planet shrinks under the assault, its density increases because the total amount of mass remains constant inside an ever-decreasing volume.

Suppose that our planet is compressed to one-quarter its present size, thus doubling the escape velocity. Anything attempting to escape from this hypothetically compressed Earth would require a velocity of some twenty kilometers per second. Imagine compressing Earth still more. Squeeze it, for example, by an additional factor of a thousand, making its radius hardly more than a kilometer. Escape from an object having a radius of about a kilometer,

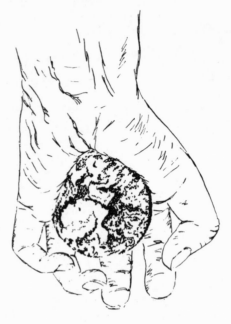

but a mass equivalent to the entire Earth, requires a velocity of many hundred kilometers per second.

Imagining our home planet being further compacted, we find the velocity of escape from this ridiculously compact "Earth" to rise accordingly. In fact, if our hypothetical vise were to compress Earth hard enough to crush its size to a centimeter across (about half an inch), then the velocity required to escape its surface would equal 300,000 kilometers per second (or 186,000 miles per second). This is no ordinary velocity; it's the velocity of light, the fastest velocity allowed by the laws of physics as we now know them.

So if, by some fantastic means, the entire planet Earth could be compressed to the size of an aspirin tablet, then the velocity required for anything to escape from it would exceed the velocity of light. And since that's impossible, the compelling conclusion is that nothing — absolutely nothing — could get away from the surface of such a compressed "Earth." There's simply no way to launch a rocket, a beam of light, or any type of radiation. Furthermore, there could be no exchange of information whatsoever with such an astronomical object. It would have become invisible and un-communicative. For all practical purposes, such a supercompact object can be said to have disappeared from the Universe!

The above example is of course hypothetical. It's unlikely that there exists anything equivalent to a vise capable of squeezing the entire Earth to centimeter dimensions. But in massive stars, such a vise does in fact exist — the force of gravity.

Gravity cannot crush Earth in this way because there's just not enough mass; the collective gravitational pull of all parts of Earth on all other parts of Earth is simply not powerful enough. However, at the end of a star's life, when the nuclear fires have ceased, gravity can literally crush a star on all sides, thereby packing an enormous amount of matter into a very small sphere.

When stars have more than several solar masses, the critical size at which the escape velocity equals that of light is not, as for Earth, in centimeter dimensions. For typically massive stellar core remnants, this critical size is on the order of kilometers. For example, a ten-solar-mass core remnant is expected to have a critical size of several tens of kilometers. To be sure, this compression is no less a feat than compressing a planet-sized object to centimeter size. In the stellar case, though, we're not talking about a hypothetical situation using an imaginary vise. The relentless pull of gravity is truly strong enough to compress entire stars to extraordinarily small dimensions. The strong gravitational force field of massive stars is not at all hypothetical; it's real.

The critical size below which astronomical objects are predicted to disappear is given a special name. Called the "event horizon," this size defines the region within which no event can ever be seen, heard, or known by anyone outside. Accordingly, the event horizons of Earth and of a ten-solar-mass star are said to be a centimeter and thirty kilometers, respectively.

We might then suggest that magicians really could make coins and other small objects disappear provided they squeezed their hands hard enough. Even people could disappear if they could arrange to be compressed to a size smaller than 10^{-23} centimeters. In English units, that's a trillionth of a trillionth of an inch. Gravity won't naturally do it to us, though. Luckily for us, we're just not massive enough. The collective gravitational pull of all the atoms in our bodies falls far short of the force required to compress us to this minuscule size. Nor is there any technological means presently known that comes close to doing so.

The important point here is the following: Should there be no force capable of withstanding the self-gravity of a dead star having several solar masses or more, then such an object will collapse naturally, and of its own accord, to an ever-diminishing size. Theory suggests that the shrinkage of a massive stellar core remnant will not even stop at a size equal to the event horizon. An event horizon is not a physical boundary of any type, just a communications barrier. The core remnant shrinks right on past it to smaller values, presumably on its way toward becoming an infinitely small point — a singularity. We say "presumably" because we cannot be sure that there isn't an as yet undiscovered force capable of halting catastrophic collapse somewhere between the event horizon and the point of singularity. The discipline required to discover any such force — quantum gravity — has yet to be invented.

Here, then, is the sequence of events expected if these late stages of stellar evolution are valid. A very massive star ends its burning cycle by exploding as a supernova. Much of the star's original content is then ejected as fast-moving debris. Provided at least a few solar masses of material remain behind, the unexploded remnant core will collapse catastrophically, the whole core diving below the event horizon in less than a second. The core simply winks out — literally disappearing — leaving a small dark region from which radiation cannot escape. This is the way a black hole is born as a blackened region of space. They are not really objects; they're just holes — black holes in space.

Black holes are strictly products of relativity theory. Whereas white-dwarf and neutron stars are valid end points of stellar evolution even within the confines of the Newtonian theory of gravity, only the Einsteinian theory of spacetime predicts the existence of black holes. As such, they are expected to obey all the standard laws of relativity theory. In particular, the mass contained within a black hole is expected to warp space and time in its vicinity. Close to the hole, the gravitational force field becomes large, the curvature of spacetime extreme. At the event horizon itself, the curvature is so great that spacetime folds over on itself, causing trapped objects to disappear.

There are several ways to visualize the curvature of spacetime near a black hole. Each way is, however, only an analogy. The problem here, as earlier in the case of the whole Universe, is our inability to deal conceptually with four dimensions.

The formation of black holes and the extreme warping of spacetime caused by them can be appreciated by imagining a large group of people living on an enormous rubber sheet — a gigantic trampoline of sorts. Deciding to hold a reunion, all except one person converge on a given location at a given time. Their reunion is to be an event in spacetime. The one person remaining behind can still keep in touch by means of balls rolled out to him along the rubber sheet. These balls are the analogue of radiation traveling at the velocity of light, while the rubber sheet is the analogue of the fabric of spacetime itself.

As the people converge, the rubber sheet begins to sag under their weight. The accumulation of the mass in a small location creates a large degree of spacetime curvature. The balls can still reach the lone person residing far away in essentially flat spacetime, but they arrive less frequently as the sheet becomes progressively more warped.

Finally, when a large enough number of people have arrived at the appointed spot, the mass becomes too great for the rubber to support. The sheet breaks, sending them into oblivion. Nor can the message balls any longer return to the lone survivor. Regardless of the speed of the last message ball, it cannot quite outrun the downward stretching sheet.

Analogously, a black hole of large mass is theorized to warp spacetime completely around on itself, thereby isolating it from the rest of the Universe.

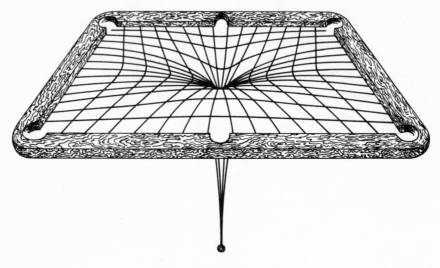

There are two important features worth noting about black holes. The first is that black holes are not cosmic vacuum cleaners; they do not cruise around interstellar space, sucking up everything in sight. The dynamics of objects near black holes are expected to mimic those of any object near a very massive star. The only difference is that, in the case of a black hole, objects orbit about a dark, invisible region. Neither emitted radiation nor reflected radiation of any sort emanates from the position of the black hole itself.

Black holes, then, do not go out of their way to drag in matter, but if some matter does happen to infall via the normal pull of gravity, it will be unable to get out. Black holes can be regarded as turnstiles, permitting matter to flow only in one direction — inward. Swallowing matter, they constantly increase their mass as well as their event horizons, for the region of invisibility also depends on the mass trapped inside. If there really are black holes in space, then they are probably enlarging their mass and size, some more than others, all of them apparently gulping, eating, growing.

A second notable point is that the strong gravity fields of black holes cause great tidal stress. An unfortunate person, falling feet first into a black hole, would find himself stretched enormously in height. He would, moreover, be literally torn apart, for gravity would be stronger at his feet than at his head. He wouldn't stay in one piece for more than a fraction of a second after passing the event horizon. Similar distortion and breakup apply to any kind of matter near a black hole. Whatever falls in — gas, people, space probes, whatever — is vertically elongated and horizontally compressed, all the while being accelerated to high speeds. The upshot is numerous and violent collisions among the torn-up debris, all of which yield a great deal of heating.

This rapid heating of matter by tides and collisions is so efficient that, prior to submersion below the hole's event horizon, newly infalling matter outside a black hole is effectively converted to heat energy while falling toward the hole. Although radiation ceases to be detectable once the hot matter dips below the event horizon, regions just outside black holes are expected to be absolutely prodigious sources of energy.

With this in mind, perhaps black hole research may result in practical applications after all. Through some marvel of technology, our descendants may someday learn how to compact garbage to an almost incredibly small size. It would then disappear! Not only that, the crushed garbage would emit copious amounts of energy in return. Maybe black holes are just what the doctor ordered for technological civilizations long on pollution and short on energy.

Of considerable interest is the obvious question, What lies within the event horizon of a black hole? What's it like deep down inside? No one knows.

Some researchers suggest that the inner workings of black holes are irrelevant. Experiments could conceivably be done to test the nature of space and time inside an event horizon, but that information could never reach the rest of us outside. It seems that no theory offered to explain the recesses of black holes can ever be put to the experimental test. Anyone's guess is as valid as anyone else's.

Perhaps the inner sanctums of black holes represent the ultimate in the unknowable. For that very reason, though, other researchers argue that it is of utmost importance to unravel the nature of black holes, lest we someday begin to worship them. Large segments of humankind have often revered the unknowable, venerating that which cannot be tested experimentally. Come to think of it, this is still the case for many of Earth's twentieth-century societies.

What sense are we to make of black holes? The equations of general relativity predict all these fantastic phenomena simply because mass warps spacetime. The larger the mass concentration, the greater the warp, and thus the weirder the observational consequences. Perhaps. Some theorists argue that relativity is incorrect, or at least incomplete, when applied to black holes. Admittedly, it doesn't seem to make much sense to argue that very massive astronomical objects collapse catastrophically to infinitely small points. Not even the wildest imaginations can visualize this phenomenon. Maybe the current laws of physics are inadequate in the vicinity of a singularity. *At* the point of singularity, we know for a fact that general relativity becomes absurd. On the other hand, perhaps matter trapped in black holes never does reach a singularity. Perhaps matter just approaches this most bizarre state, in which case relativity theory may still hold true. Scientists just don't know yet.

What's the story? Do black holes really exist? Or are they nothing more than figments of theorists' fertile imaginations? Maybe all massive stars are blasted to smithereens when they explode as supernovae, never leaving much of a remnant core at all. Or perhaps there is another, as yet undiscovered force capable of competing with gravity despite these extreme conditions of ultracondensed matter. Each of these possibilities would preclude the existence of black holes. Just how much observational evidence is there for black holes?

Despite the fact that black holes are invisible, they are expected to remain intense sources of gravity. Accordingly, astronomers can test for their existence by examining the gravitational force field throughout the space near them. For example, the gravitational response of a spacecraft or of a nearby celestial body could conceivably be used to probe the nature of a black hole. Objects outside an event horizon are predicted to behave just as though there were a massive, visible object at the site of the hole. In other words, all conventional evidence of a black hole disappears, but its gravity persists.

Well, our civilization doesn't have the capability to maneuver spacecraft into the neighborhood of suspected black holes, even if we knew their exact locations. But the Galaxy is known to house numerous double-star systems whose members orbit about one another. Yet, for many of these, only one star is visible. Of course, each unseen companion could be just a small and dim star, hidden in the glare of a large and bright stellar partner. Or the object could be shrouded in dust or other interstellar debris, making it invisible to equipment on Earth, but not necessarily indicative of a black hole. Indeed, this is probably the case for the majority of binary star systems having an unrevealed member, in which case the invisible candidates are not black holes.

A few of these peculiar binary systems, however, display properties suggestive of black holes. The most interesting discoveries, made only in the 1970s, involve those binary systems that emit copious amounts of X-ray radiation. This high-frequency radiation cannot easily penetrate dust, rendering unlikely the hypothesis that galactic debris has camouflaged one of the partners. Indeed, several Earth-orbiting satellites equipped to detect X rays have recently observed just this type of radiation from some binary systems. And in several of these cases, only one star can be seen visually.

Observations have shown that, in each case, the X-ray radiation arises from billion-degree-Celsius gas flowing from a large visible star toward a small, unseen companion. Furthermore, each invisible region encompasses five to ten solar masses, and spans no more than a hundred kilometers. They have all the earmarks of black holes.

A tentative model for the nearest such system, some six thousand light-years from Earth, stipulates that much of the gas drawn from the visible star will end up in a flattened Life-Saver-shaped disk of

matter. Some of this gas inevitably streams toward the black hole and becomes trapped by it.

Several problems plague these interpretations, not the least of which is that all the suspected black holes in binary systems have masses close to the neutron star–black hole dividing line. When the effects of rotation and magnetism are incorporated into the theory, there's a chance that the dark objects in question may come to be recognized as very dim neutron stars, and not black holes at all.

Many researchers argue that other kinds of regions display better evidence for candidacy as black holes. Current exploration of the center of our Milky Way Galaxy, some thirty thousand light-years from Earth, is especially interesting in this regard. Although the middle realms of our Galaxy are totally obscured by dust and thus cannot be studied with optical telescopes, radio and infrared observations of the innermost few hundred light-years have yielded spectacularly unexpected results. Data obtained only in the late 1970s imply the presence of rapidly rotating hot gas, suggestive of a colossal whirlpool at the very center of our Galaxy.

A model capable of accounting for the observations stipulates that a halo of thin, hot, ionized gas surrounds a core of even hotter, denser gas. The whole swirling region is hypothesized to orbit — and this is the punch line — a region having a mass equal to several million suns, though a size hardly larger than our Solar System. Such an enormously massive and compact blob is required to keep the whirling gas from dispersing into the outer regions of the Galaxy; fast rotation rates doubtless produce strong centrifugal forces and, unless there were a huge mass gravitationally pulling back, the gas would be flung away like mud from the edge of a spinning bicycle wheel.

Though the details are controversial, a consensus now seems to be emerging that a supermassive but ultracompact "something" resides at the very heart of our Milky Way Galaxy. It seems that that something can be only one thing — a black hole in space.

Recent observations suggest that supermassive objects also lurk in or near the central regions of a few other nearby galaxies. The evidence here is much the same as for our own Galaxy, with gas and stars in the innermost region of several active galaxies observed to be rapidly whirling. Detailed measurements of the active galaxies furthermore imply a compact region harboring even greater mass than in the case of our normal Galaxy. In fact, several billion solar masses are inferred. Perhaps these central whirlpools are remnants of the turbulent eddies that gave rise to the galaxies in an earlier epoch, as discussed previously.

We may find that the center of every galaxy is inhabited by a supermassive black hole. Normal galaxies such as our own probably have relatively small black holes of "only" millions of solar masses. More active galaxies would be expected to have progressively larger black holes, perhaps on the order of billions of solar masses. Don't forget that, in the process of gobbling matter, black holes release vast amounts of energy; gravitational infall of large quantities of matter toward the hole accelerates and heats the matter, causing it to radiate much of this energy before disappearing below the event horizon. The great energetics and explosiveness of the active galaxies might be naturally explained by matter perishing within the clutches of supermassive black holes.

It's even possible that the most energetic objects in the Universe — the quasars — could be powered by hypermassive black holes that swallow whole stars. Although no observational evidence currently exists to support it, this idea is a mere extension of the above arguments. If true, then black holes in quasars would presumably be even more massive, more compact, and more bizarre than the billion-solar-mass objects suggested for the active galaxies.

Clearly, as noted earlier, an understanding of the powerhouse galaxies lies buried within their hearts, awaiting future astrophysical explorers to discover, unravel, and share their secrets.

A final word of caution is in order concerning black holes. Force fields may yet be discovered capable of withstanding the relentless pull of gravity, even that near exceedingly massive and compact astronomical objects. Magnetism and rotation have not yet been fully incorporated into the theory of black holes, and no one knows what to expect regarding the behavior of gravity on microscopic scales. To be sure, these are terribly hard problems, so tricky that some of the best minds confess ignorance as to how to go about even attacking them. Indeed, serious research regarding realistic models of black holes is only beginning at many observatories around the world.

Unless astrophysicists can find direct, or undeniable indirect, evidence for the existence of black holes, neither of which is currently at hand, then the whole concept of black holes may well turn out to be no more than a whim of human fantasy — another myth to be held guilty until proven innocent. The nature of matter, energy, space, and time deep down inside event horizons may be no more significant than challenging and amusing academic problems devoid of reality.

On the other hand, the Universe did arise from what would seem to have been a bare singularity more than ten billion years ago. Of all the material coagulations now known or suspected to be part of our Universal inventory, black hole singularities might just be the keys needed to unlock an understanding of the ultimate singularity. By theoretically studying the nature of black holes, and especially by experimentally seeking their existence and probing their behavior, we may someday be in a better position to address *the* most fundamental problem of all — the origin of the Universe itself.

PLANETS

Habitats for Life

billions of years ago

PLANETS ARE GLOBES OF MATTER, much smaller than stars, and largely composed of heavy elements. These worlds could not have formed early in the Universe. There simply were no heavy elements early on.

Planets had to await the birth and death of countless high-mass stars, which seem to be the only locales suitable for the creation of heavy elements. The act of stellar death ensures a recurring fertilization of interstellar space, from which later-generation stars as well as planets can originate. Planets are, then, quite literally, collections of the cinders of burned-out stars.

The planetary group in which we live is a varied lot. Known as the Solar System, it includes one star, nine planets, at least three-dozen moons, thousands of asteroids ranging in diameter from one to a few hundred kilometers, myriads of comets of kilometer dimensions, and innumerable interplanetary meteoroids less than a meter across. With the Earth-Sun distance of about a hundred million kilometers termed an astronomical unit, the entire Solar System extends end to end for nearly eighty astronomical units. That may sound large, but it's only about a thousandth of a light-year.

The four innermost planets, Mercury, Venus, Earth, and Mars, are often termed the Terrestrial Planets because of their physical and chemical similarity to rocky Earth. The larger, outer planets, Jupiter, Saturn, Uranus, and Neptune, are often labeled the Jovian Planets because of their resemblance to gassy Jupiter. Between these two groups, in a well-defined belt some two to three astronomical units from the Sun, roam the stony asteroids, sometimes called "minor planets" or even "planetoids." Pluto, usually the outermost planet (though it sometimes ventures inside Neptune's orbit, as it did in 1979), doesn't fit well into any of these categories; it was probably once a moon of Neptune, not originally a planet at all.

From a distant vantage point, the Sun dominates, with Jupiter an inferior second. Our star has about a thousand times Jupiter's mass, and about seven hundred times that of the whole rest of the Solar System including Jupiter. The Sun, then, contains more than 99.9 percent of all the matter in the Solar System. Everything else, especially the small Terrestrial Planets and notably Earth, resembles a collection of nearly insignificant debris.

Draw a distinction in Jupiter's case, however, for this is no ordinary heavenly body. Jupiter in fact just missed becoming a star. The composition and structure of this giant planet — and possibly all the Jovian Planets — is stellar. But none of them is big enough to ignite. If Jupiter were only a hundred times as massive, its central temperature would just about equal that necessary to commence nuclear reactions, converting it into a small star. Thus, our Solar System almost formed as a binary star system, an astronomical posture that would have undoubtedly rendered Earth life impossible.

We owe a debt of gratitude to the Sun for lighting up, and to Jupiter for not.

The earlier apparent complexity of the many objects in our Solar System was greatly simplified in Renaissance times. Looking (observation) and thinking (theory) were amalgamated to achieve a more objective view than that deduced by the ancients. The sixteenth-century Polish cleric Nicholas Copernicus recognized

that a heliocentric (Sun-centered) model improved the harmony of the tangled geocentric (Earth-centered) models imagined by the Greeks and Romans.

Despite the support of observational data and a mathematical underpinning by two seventeenth-century giants, Kepler and Newton, the simpler model of Copernicus was not easy to accept even as recently as three hundred years ago. Heliocentricity rubbed against the grain of all previous thinking. And it violated the religious teachings of the time. Above all, it relegated Earth to a noncentral and undistinguished location within the Solar System and the Universe. Earth became just one of several planets.

Although we now recognize that these Renaissance workers were correct, none of them was able to prove to his contemporaries that our system is centered on the Sun, or even that the Earth moves. Proof of the latter came early in the nineteenth century when the first stellar parallax observations were made by the German astronomer-mathematician Friedrich Bessel. Heliocentricity of the Solar System has been verified gradually over the years with an ever-increasing number of experimental tests, culminating with the expeditions of our unmanned space probes of the last two decades.

The initial motivation for the heliocentric model was simplicity. It provides a more natural explanation of the observed facts than can any geocentric model. Even today, scientists are guided by simplicity, symmetry, and beauty in modeling all aspects of the Universe.

Development and eventual acceptance of the heliocentric model is a prominent milestone in our thinking as human beings. Fathoming the framework of our planetary system freed us from an Earth-centered view of the Universe, and enabled us to realize that Earth orbits only one of myriads of similar stars in the Milky Way Galaxy. Surprisingly enough, it was hardly more than a half-century ago that the American astronomer Harlow Shapley took the next bold step, declaring that neither was our Sun centralized, unique, or special in any way. The more we look and the more we test, the more mediocre our niche in the Universe seems to be.

∞

Any model capable of explaining the origin and architecture of the planets and their moons must adhere to the known facts. Generally, these facts emerge from studies of interstellar gas clouds, meteorites, and Earth's Moon, as well as from the various planets now observed with ground-based telescopes and interplanetary space probes. The meteorites provide especially useful information, for they have entrapped traces of solid and gaseous matter uneroded from the early Solar System. Radioactive dating of all meteorites uniformly suggests that the Solar System formed, with the Sun and Earth as part of it, approximately five billion years ago. Studies of terrestrial and lunar rocks generally confirm this date.

Among the many observed properties of our Solar System, seven stand out. They can be summarized as follows:

Each planet is relatively isolated in space, none of them being bunched together; each planet resides roughly twice as far from the Sun as its next inward neighbor.

The orbits of the planets nearly describe circles; the innermost planet Mercury's slight orbital eccentricity is caused by great tidal stresses exerted on it by the neighboring Sun, while that of Pluto can be explained by hypothesizing that this outermost planet is an escaped moon of the planet Neptune.

The orbits of the planets nearly all lie in the same plane; each of the planes swept out by the planets' orbits aligns with the others to within a few degrees, the whole system taking on the shape of a rather thin disk.

The direction in which planets revolve in their orbits about the

Sun is the same in which the Sun rotates on its axis; virtually all the angular momentum in the Solar System — the planets' orbits and the Sun's spin — seems to be systematized, suggesting a high degree of unison.

The direction in which the planets rotate on their axes also mimics that of the Sun's spin; the two exceptions are Venus, which appears to spin in the opposite way, and Uranus, whose poles seem to lie in the plane of that planet's orbit.

Most of the known moons revolve about their parent planets in the same direction that the planets rotate on their axes; some moons, like those associated with Jupiter, seem to form miniature Solar Systems, revolving about their parent planet in roughly the same plane as the planet's equator, and again suggesting a high degree of unison throughout all aspects of our planetary system.

Finally, the Solar System is highly differentiated; the inner, Terrestrial Planets are characterized by high densities, moderate atmospheres, slow rotation rates, and few or no moons, while the outer, Jovian Planets have low densities, thick atmospheres, rapid rotation rates, and many moons.

All these observed facts, when taken together, strongly suggest that there is a large degree of order within our Solar System. The whole ensemble is apparently not a random assortment of objects spinnning or orbiting this way or that. It hardly seems possible that the Solar System is a pickup team, amassed by the slow accumulation of already-fashioned interstellar "planets" casually captured by our Sun over the course of billions of years. The overall architecture of our Solar System seems too neat, and the ages of its members too uniform, to be the result of a long series of haphazard circumstances. All signs point toward a single formation, the product of an ancient but one-time event.

Though not all these properties were recognized two centuries ago, the crux of the modern theory of our Solar System's formation dates back that far. Known as the condensation hypothesis, its origin is often attributed to Immanuel Kant, but he merely elaborated upon an earlier proposal made in the seventeenth century by the French philosopher René Descartes. In this model, a giant swirling region of matter is visualized as giving rise to planets and their moons as a natural by-product of the star formation process. But these philosophers failed to work out the mathematical details of their model; their proposals amounted to little more than qualitative ideas.

Later in the eighteenth century, a French mathematician-astronomer named Pierre Simon de Laplace attempted to furnish this type of model with a quantitative basis. He was able to show mathematically that angular momentum arguments demand that the contracting matter of an interstellar fragment spin faster; a decrease in the size of a rotating mass must be balanced by an increase in its rotational speed, much like a pirouetting figure skater who spins faster while bringing in her arms, or a high diver who somersaults quickly by tightly curling his body. The fragment eventually flattens into a pancake-shaped primitive Solar System, for the simple reason that gravity can pull matter toward the center of the region more easily along the rotation axis than perpendicular to it. This model provides a plausible origin of some of the ordered architecture observed in our Solar System today — the circularity of the planets' orbits, their coplanar distribution, and many of the other properties noted above. These are mere products of the natural changes experienced by an interstellar fragment, a straightforward obedience of a parcel of galactic gas to the known laws of physics.

Continued contraction of such a primitive Solar System forces the entire mass to spin more rapidly as time progresses. Near the fringe, the outward centrifugal push eventually exceeds the inward gravitational pull. The push creates a flattened ring of gaseous matter which breaks away from the rest of the primitive Solar System. The rest of the system then contracts a little more until such time as another ring of matter is deposited inward of the first. Progressing in this way, an entire series of rings can be fashioned about the central protosun. Each ring is furthermore theorized to

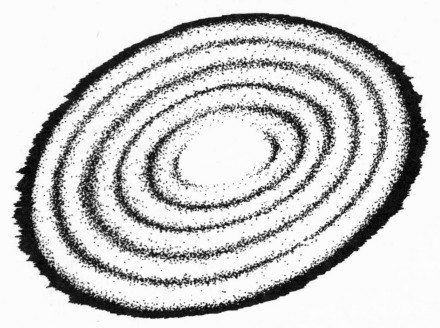

condense, over long intervals of time, into a planet. Several outer planets might develop while the interior of the primitive Solar System continued to shape the inner planets and the Sun.

As nice as this condensation hypothesis may seem, it's not without difficulties. Detailed computations show that material in a ring of this sort would not necessarily condense to form a planet. In fact, recent computer calculations predict just the opposite; the rings would tend to disperse, owing to both a wealth of heat and a lack of mass within any one ring. Gravitational coagulation of interstellar matter is one thing; it works reasonably well when fabricating stars because, after all, there's a vast amount of mass contained within a typical interstellar cloud. But coagulation of a warm, protoplanetary ring is another thing; there's not nearly enough matter for gravity to gather it into a planet-sized ball. Instead of coalescing to form a planet, the computations predict the ring will break up and fade away.

Don't be too hard on Laplace. He didn't have a computer, and it's both tricky and tedious to account for all the statistical subtleties of this problem without one.

There's a second problem with the condensation hypothesis for the origin of the Solar System. The Sun spins on its axis once in about thirty days, a good deal more slowly than Earth, which rotates once in twenty-four hours. This solar sluggishness baffles the experts. Why? Although the Sun claims nearly all the matter in the Solar System, it possesses only about two percent of the momentum of the entire system. Jupiter, for instance, has a lot more momentum than our Sun. The planet does not spin on its axis overmuch, but an object with Jupiter's considerable mass so distant from the Sun carries a great deal of momentum in its orbit. In fact, Jupiter harbors more than half of the Solar System's present momentum. All told, the four big Jovian Planets account for approximately ninety-eight percent of the present momentum of the Solar System. By comparison, the lighter Terrestrial Planets have negligible momentum.

The condensation hypothesis predicts that the Sun should command most of the Solar System's momentum. After all, the Sun has most of the mass. Why then shouldn't it have most of the momentum? This is especially true since contracted objects are expected to increase their spin rate, again in the manner of the figure skater's pirouette. Expressed another way: If all the planets, with their large amounts of orbital momentum, were hypothetically deposited inside the Sun, it would spin on its axis a hundred times as fast as at present. Instead of rotating about once a month, the Sun would spin once every several hours.

These and other difficulties with the condensation hypothesis forced researchers to consider alternative models. One such model is called the encounter hypothesis, or, more popularly, the collision hypothesis. In it, the planets are visualized as the condensed products of hot, streaming debris torn from our Sun during a close encounter with another star. The flaming streamers produced by such a near collision are then surmised to remain gravitationally bound to our Sun, to be captured furthermore into orbits about

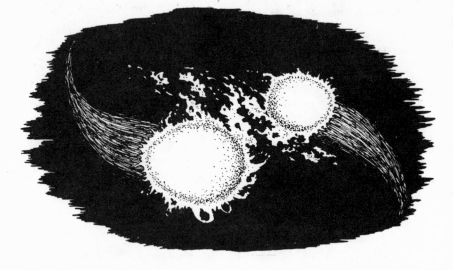

it, and to condense eventually into planets. Despite the phenomenal tides that would undoubtedly accompany the near collision of two stars, the predicted aftermath is consistent with the common orientation of the planets' orbits and the Sun's spin, as well as the close planar alignment of all the planets.

Although this model was also first proposed during the eighteenth century, modern enthusiasm for it was rekindled about a hundred years ago when it became clear that there were a few minor exceptions to the ordered architecture of our planetary system. The absolute simplicity of the Solar System broke down, giving a boost to catastrophic models — models that invoke accidental and unlikely celestial events.

Though the encounter hypothesis has some qualitative points in its favor, it also has its share of pitfalls. The high improbability of a near collision between two stars is the foremost problem. Stars are large by terrestrial standards, but still minute compared to the distances that separate them. For example, the Sun is about a million kilometers in diameter, whereas the distance to Alpha Centauri, the nearest star system, is almost a hundred trillion kilometers. Probability theory then suggests that, given the number of stars, their sizes, and their typical separations, there could have been no more than a handful of such near collisions throughout the entire expanse and history of the Milky Way Galaxy. Galaxy collisions are frequent, but stellar collisions are extremely rare.

The improbability of such a collision does not, of course, prove that the encounter hypothesis is wrong. After all, our Solar System could be the foremost — even the only — example of this extraordinarily uncommon phenomenon. Should this hypothesis be correct, we can justifiably conclude that our planetary system is a rare type of astronomical region. Very few stars would be expected to have planets, and the chances for extraterrestrial life diminish accordingly.

Besides the small chance of collision, several other difficulties plague the idea of planetary origin via encounters of any kind. First of all, the momentum puzzle besetting the condensation hypothesis is also a problem here. Second and more formidable, it's hard to understand how hot material torn from the Sun could contract; hot gases usually disperse. Consequently, although such a near collision between two stars might occasionally happen, it's unlikely that the hot fragments would form planets. Some of the hot streamers would surely fall back into the Sun. Others would tend to disperse more quickly than the matter in the cooler rings of the condensation hypothesis. A third quandary concerns the nearly circular orbits traced by each of the observed planets. If matter were tidally ripped from the Sun to form the planets, why should each of the clumps of debris end up orbiting that Sun in a near-perfect circle? The encounter hypothesis cannot explain this observed fact, even qualitatively.

Yet another model of how the Solar System formed, the one embraced by most astrophysicists, is termed the nebular hypothesis. It's really a souped-up version of the condensation hypothesis discussed earlier. It mixes all the attractive features of the old condensation hypothesis with our recently revised assessment of interstellar chemistry. Theorists can now concoct a nebular hypothesis that alleviates several of the aforementioned problems.

Recall that the first problem with the condensation hypothesis concerned the inability of the ringed material to coagulate into a tight-knit ball of protoplanetary matter; each ring was expected to have too little mass and too much heat to initiate gravitational contraction. However, a new twist has been added only within the past decade or so. Astrophysicists have come to realize the ubiquity and importance of dust grains in interstellar space.

Dust grains, solid microscopic bodies of rock and ice, play a crucial role in the evolution of any gas. The nebular model assumes that dust, peppered throughout the warm gas of the primitive Solar System, helps to cool it by efficiently radiating heat away from protoplanetary blobs. Furthermore, dust grains enhance the coagulation of atoms within the gas, acting as condensation nuclei around which other atoms can aggregate. The presence of dust guarantees that the gaseous matter coagulates by cooling below the point at which pressure pushing out can effectively compete with gravity pulling in.

By postulating the existence of a dusty interstellar cloud five billion years ago, theorists now predict dust-grain cooling to have occurred before the gas had a chance to float away. In this way, modern observations of sooty interstellar matter suggest, but do not prove, the likelihood of assembly rather than dispersal of protoplanetary matter.

To trace the formative stages of a planetary system such as ours, modern nebular models stipulate the following broad scenario. Imagine a dusty interstellar cloud fragment measuring about a light-year across. Intermingled with the usual plenitude of hydrogen and helium atoms, the cloud harbors some heavy-element gas and dust, an accumulation of ejected matter from many past supernovae. Gravitational instabilities start the fragment contracting to the size of about a hundred astronomical units, after which dense protoplanetary eddies form of their own accord.

These instabilities could have been caused by many events, but the view now favored by the astronomical community is that another supernova was probably the culprit. Anomalies in the

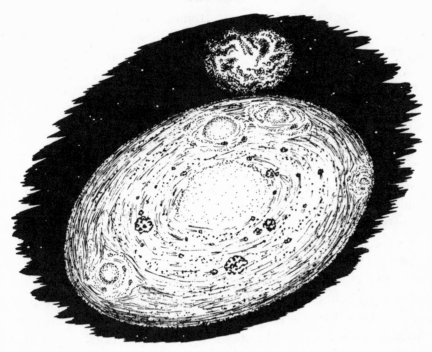

elemental abundances of captured meteorites suggest that the formation of our Solar System might have been triggered by the concussion of a nearby supernova that popped off some five billion years ago. The explosion would have sent a shock wave through the interstellar cloud, piling up matter into dense sheets, much like snow swept by the blade of a plow.

Calculations show that when a shock wave encounters an interstellar cloud, it races around the thinner exterior of the cloud more rapidly than it can penetrate into its thicker interior. Shock waves do not blast a cloud from only one direction; they squeeze it from many directions. Atomic bomb tests have experimentally demonstrated this squeezing: Shock waves created in the blast surround buildings, causing them to be blown together (imploded), rather than apart (exploded). Similarly, shock waves can cause the initial compression of an interstellar cloud, after which natural gravitational instabilities divide it into fragments which gradually form stars and planets. Ironically, the demise of old stars may be the trigger necessary to conceive new ones.

Once the shock wave passed, turbulent gas eddies would appear naturally at various locations throughout the primitive, rotating Solar System, the bulk of which, by this time, would have flattened into a Frisbee-shaped disk. As in the earlier cases of galaxy and star formation, these eddies would be nothing more than gas density fluctuations that come and go at random. Provided an eddy was able to sweep up enough matter while orbiting the protosun, including a rich enough mixture of dust to cool it, then gravity alone would ensure the formation of a planet. The entire process can be likened to a snowball thrown through a fierce snowstorm, growing larger while encountering more snowflakes. In this way, individual planets could be fabricated by accreting most of the cold matter at varying distances from the protosun. Assuming the "sweeping" process was reasonably efficient throughout the disk, we can begin to understand how our present Solar System has come to exist as a collection of rather small planets orbiting throughout an otherwise empty region of space.

The natural satellites or moons of the planets presumably formed in similar fashion, as still smaller eddies condensed in the vicinity of their parent planets.

Mathematical modeling suggests that a hundred million years would have been needed for the primitive Solar System to evolve nine protoplanetary eddies, dozens of protomoons, as well as the big protosolar eddy at the center. It's estimated that another billion years would have been required to sweep the system reasonably clear of interplanetary trash.

The weakest link in the nebular hypothesis is, once again, the anomalously small momentum of our present Sun. All mathematical modeling requires the Sun to have been spinning very fast in the earliest epochs of the Solar System. Somehow, it must have lost virtually all its spin, but no consensus has thus far been reached on how it managed to do so.

Some researchers speculate that the solar "wind," discovered only in the 1960s by automated spacecraft, could be the root of the problem. High-velocity elementary particles constantly escaping the Sun through flares and other surface storms could act to break the Sun's spin. Particles boiled off the Sun must carry with them minute amounts of the Sun's momentum and, over the course of five billion years, could have robbed the Sun of much of its initial momentum. Manned and unmanned space vehicles are now attempting to measure the intensity of solar activity, though it will be most difficult even to estimate the level of that activity billions of years ago.

Other researchers prefer to solve the Sun's momentum problem by postulating a primitive Solar System considerably more massive than the present-day system. They argue that the accretion process was not entirely successful during the system's formative stages. Matter not captured by the Sun or the planets may well have transported some momentum while escaping back into interstellar space. This proposal is tough to test, since the escaped matter would be well beyond the range of our current robot space probes.

Despite some controversy as to how this momentum quandary can best be resolved, nearly all astrophysicists agree that some version of the nebular hypothesis is correct. The details, however, not yet worked out, form the essence of a most troubling problem now being addressed at several observatories around the world.

∞

Diversity of physical conditions in the earliest years of the Solar System is probably responsible for the large contrast between the Terrestrial and the Jovian Planets. As the primitive Solar System contracted as a result of gravity, it heated and flattened. Dust grains broke apart into molecules, and they in turn into atoms. Since the density, and hence the collision rate, was surely greater close to the protosun, material there would have become hotter than in the outlying portions of the primitive Solar System. Calculations suggest that while the gas temperature was several thousand degrees Celsius near the core of the contracting system, it would have been only several hundred degrees Celsius some ten astronomical units away, out where Saturn now resides.

Such a gas cannot continue to heat indefinitely, lest the region blow up. Like any hot gas, the primitive Solar System must have released some of its newly gained energy. So, even as the protosun continued heating upon contraction, the outer regions of the primordial system cooled. As a result, heavy elements several astronomical units from the center of the protosun must have crystallized from their hotter gas phase to their cooler solid phase. (The same process occurs today, though on a smaller scale, on Earth, as raindrops, snowflakes, and hailstones condense from moist, cooling air.)

With a further passage of time, the temperature decreased at all locations, except at the very core where the Sun was forming. Everywhere beyond the protosun, atoms returned to their unexcited states, after which some of them collided and stuck to form molecules, which in turn coagulated to form dust grains once more.

You might think it amusing that, although there were plenty of interstellar dust grains early on, nature saw fit to destroy them only to rebuild them again later. However, an important change occurred in the process. Initially, the interstellar gas was evenly peppered with an array of all sorts of dust grains. When the dust later reformed, the mixture was much different, for the condensation of solid dust from hot gas depends on the temperature. In other words, the act of contractive heating served to sterilize the entire region, thus setting the stage for a Solar System highly diversified in planetary composition.

In the outer regions of the primitive planetary system, beyond several astronomical units from the center, the temperature would have been right for the condensation of several abundant gases into solids. At temperatures of two hundred degrees Celsius or less, the reasonably abundant heavy elements such as oxygen, carbon, and nitrogen combined with the most abundant element, hydrogen, to form some well-known simple chemicals, including water, ammonia, and methane ice crystals. (Helium is an inert element and does not combine chemically with other atoms.) The ancestral

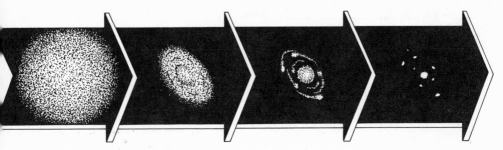

fragments destined to become the Jovian Planets were formed under rather cold conditions by gravitational instabilities much like those discussed earlier for the formation of galaxies and stars.

Microscopic icy grains orbiting throughout the nebular disk gradually collided and stuck together, fabricating increasingly larger aggregates of ice in much the same way that fluffy snowflakes can be compressed into snowballs. Together with leftover hydrogen and helium atoms trapped by the strong gravitational pull of these protoplanets, gassy and icy compounds are now known to comprise the bulk of the Jovian Planets. There is little doubt that these massive planets formed in much the same way as our Sun. None of them is quite massive enough, though, to kindle nuclear burning, the hallmark of any star.

In the inner regions of the primitive Solar System, the average temperature would have been about a thousand degrees Celsius at the time when condensation from gas to solid began. The environment there was simply too hot for ices to survive. Instead, many of the abundant heavier elements such as silicon, iron, magnesium, aluminum — the heavy metals — would have combined with oxygen in order to produce iron oxides, silicates, and a variety of other rocky minerals.

These orbiting rocky grains gradually coalesced into objects of pebble size, baseball size, basketball size, and larger. The bigger they grew, the quicker gravity helped them to coalesce, sweeping more and more matter from the surrounding regions of the flattened nebular disk, and eventually fabricating planet-sized objects.

The bulk of the formation process probably took not much longer than a few million years. Such accretion of the condensed solids then explains the composition of the Terrestrial Planets. The very abundant light elements of hydrogen and helium, as well as many other gases that failed to condense into solids, would have surely escaped from these small protoplanetary objects. The temperature was too high, and the gravity too low, to prevent gases from escaping the inner planets. What little hydrogen and helium did manage to stick around was probably blown away by the wind and radiation of the newly formed Sun. What remained, so say the theoretical models, was a few rocky planets, each cool, hostile, and largely devoid of an atmosphere.

Why the myriad rocks of the asteroid belt between Mars and Jupiter failed to coalesce into a planet remains a mystery. Perhaps one did exist, after which it blew up, the puzzle then being for what reason.

The origin of Earth's Moon is also uncertain. Generally, three theories have been advanced to account for it. One theory suggests that the Moon condensed as a separate object near the Earth, and in much the same way as did our planet. The two objects then essentially formed a binary planet system, one revolving about the other. Though favored by many planetologists, this idea suffers from a major flaw: the Moon has a lower density and a different composition from Earth's, making it hard to understand how both could have originated from the same protoplanetary blob.

A second theory maintains that the Moon condensed far from Earth, and was then later captured by it. In this way, the density and composition of the two objects need not be similar, for the Moon presumably materialized in an alien region of the early Solar System. However, the objection here is that the Moon's capture would not be a trivial event; it's more likely an impossible one. Why? Because the mass of our Moon relative to that of Earth is larger than for any other moon of any other planet. It's not that our Moon is the largest natural satellite in the Solar System; but it is unusually big compared to its parent planet, the Earth. Mathematical modeling suggests that it's unreasonable to expect Earth's gravity to have attracted the Moon in just the right way to capture it during a close encounter sometime in the past.

A third hypothesis stipulates that the Moon originated out of the Earth itself. The Pacific Basin has often been mentioned as the place from which protolunar material may have been torn, the result of centrifugal forces on a young and rapidly rotating Earth. As absurd as this idea may seem, the early findings of the Apollo Program seemed to favor it. Both the lunar composition and density were found to mimic those of Earth's mantle, that region just below the crust. However, recent, more exacting studies of our Moon's composition show significant dissimilarities to Earth's underbelly. Also, there remains the fundamental mystery of how Earth could possibly have ejected an object as large as our Moon.

Clearly, none of these theories is satisfactory. Each suffers from a major flaw. Yet, one of them or some version of them, it would seem, must be correct. Unfortunately, direct samples of lunar material collected during the American and Russian missions have failed to settle one of humankind's most ancient questions. Perhaps the formation of our Moon was the product of circumstances so rare that we will never be able to unravel the details of its birth. Indeed, the origin of Earth's Moon is a frustrating subject — so perplexing that some researchers have been forced, in desperation, to suggest that the Moon does not exist.

At any rate, the current consensus is that the genesis of a planetary system like our own is a natural, probably frequent, outgrowth of the birth of a star. Precisely how the small atoms of gas and grains of dust managed to coalesce into the planets and moons now seen remains, however, one of the great unsolved problems of science. It's most troubling because of our inability to test the geological record of the first half-billion years of Earth's history. Material from this critical time domain, which would ordinarily provide clues to the environment in which our planet was born, is missing, having been literally melted and eroded away long ago. The only way to recover those clues to whatever did happen here approximately five billion years ago is to engage in an active program of exploration throughout our Solar System.

∞

One further issue is worth noting before we complete our discussion of planets. The leading hypothesis for the origin of the Solar System holds that the events that produced it are not at all unique. Many stars are expected to have a planetary system of some sort. Even if only one percent of all the stars in the Galaxy have planetary systems, that still leaves billions of stars with planets. And each star, of course, would probably have more than a single planet associated with it.

Theory is one thing, but observation is another. Is there any experimental evidence for planets circling other stars? Unfortunately, the answer is currently ambiguous. Certainly there is no proof that Earth-like planets are orbiting any other star. The light reflected by any planet orbiting even one of the closest stars would be too faint to detect with the very best equipment now employed by astronomers. Possibly sometime in the 1980s, when large optical telescopes are launched into orbit above Earth's blurry atmosphere, astronomers may be able to detect some Jupiter-sized planets circling stars near our Solar System. But even then, such stars would be expected to be at the threshold of visibility, and the experiments might not produce fruitful results. The images of any planets orbiting those starry points of light in the nighttime sky are bound to be puny, pale, and puzzling.

There are two types of *indirect* evidence that planets may be orbiting a few of the nearest low-mass stars. The first piece of evidence concerns the fact that more massive stars generally spin a lot faster than their low-mass counterparts. Stars having less than twice the mass of our Sun have a good deal less rotational momentum than expected. Like the Sun, most low-mass stars seem to have lost most of their original spin. Their momentum may instead be wrapped up in the orbital motions of associated planets. After all, the planets of our Solar System account for most of its angular momentum, prompting speculation that most low-mass stars have planets as well.

The second bit of indirect evidence for the existence of planets near other stars is derivable from the small gravitational influence that planets are expected to have on stars. A large planet orbiting about a low-mass star should produce a slight alteration in that star's motion. Even though the planet itself cannot be seen, such an invisible object pulls gravitationally first one way and then the other during its annual orbit, the result being a slight back-and-forth shift or wobble in the path of the star. There have been recent reports of a few stars having such a wobble, but the measurements are extraordinarily hard to make with any degree of certainty.

It seems fair to say that astronomers presently have no direct evidence, and little indirect evidence, for the existence of planets associated with any other star. For all we know, the Universe could be teeming with rocky and icy planets or even basketball-sized objects near or far from every star. Our civilization has not yet invented the equipment necessary to make an inventory of small, compact, dark clinkers residing in even nearby space.

We are left with the notion that, if planets do form as natural by-products of star formation, then space must be infested with them, just as it bristles with stars in our Galaxy, and galaxies beyond. But the feeling is an uncertain one, for we don't know for sure.

∞

Our Solar System harbors a vast array of material objects. Planets, moons, asteroids, comets, and meteoroids are all well-known, if not yet well-understood, inhabitants of our minuscule niche in the periphery of the Milky Way. The wide range of physical and chemical properties among the peculiar planets and their odd moons yields the impression that our Solar System is full of debris — scattered remains of a more violent, but at the same time more formative, era in the history of our local environment. Can we ever hope to identify all the pieces and understand the whole puzzle? The answer, we think, is yes.

Every time a new discovery is made, we learn a little more about general planetary properties. Interplanetary space probes of the past decade have made especially startling advances — in some cases forcing us to revise completely our perceptions of some astronomical objects. It's becoming clear that each planet contributes to the total appreciation of a planetary life cycle just as different stars add to our understanding of the stellar life cycle. Each offers something about the system's origin, and something about its evolution.

Almost every planet or moon is now in a different stage of development, much as red giants and white dwarfs represent varied stages of stellar evolution. The Jovian Planets resemble interstellar fragments frozen in time, not massive enough to become stars, and yet too massive to develop into huge rocks. They have preserved in varying degrees the pristine properties of the primitive Solar System. The less massive Terrestrial Planets, on the other hand, have evolved a great deal, generating hard surfaces, while outgassing atmospheres and sometimes oceans. Significantly, at least one of these small planets has spawned life.

Each planet has apparently evolved to an extent that depends on its mass. Mass does indeed play a crucial role in the evolution of all types of matter, including galaxies, stars, planets, and life. It may be the most important single developmental factor anywhere.

View the Solar System, then, as not just a collection of planetary debris. Every planet has something to tell us. Each time a new space probe reconnoiters a planet — and some probes are on their way right now — we learn a little more about that planet. In particular, we learn how that planet fits into the general scheme of the Solar System.

Eventually, a synthesis of all planetary studies should provide nothing less than a complete deciphering of the origin, evolution, and destiny of our tiny home within this vast Universe.

LIFE

Matter plus Energy

billions of years ago

WE OWE THE EXISTENCE of not only our planet but also ourselves to ancient supernovae, for we too are made of heavy elements. Literally, many stars have died so that we might live.

Nearly everything on Earth is composed of elements heavier than hydrogen and helium. The bluish oceans of liquid water consist partly of a heavy element, for water is after all not only two parts hydrogen but also one part oxygen. Only a fortuitous combination of temperature and pressure, unlike that on any other known planet, allows large quantities of water, enough to cover three-quarters of our planet's surface, to remain in the liquid phase.

Air, too, is made of heavy elements. Composed primarily of nitrogen, oxygen, and water vapor, the atmosphere is constantly subjected to meteorological change, causing the variable cloud patterns that dominate our planet from afar.

Additional heavy elements can be found in the brown and gray tracts of soil and rocky land. Looking immutable from space, and feeling stable beneath our feet, silica-rich plates of Earth's surface are nonetheless drifting across our globe, gliding at an imperceptibly sluggish pace.

* * *

147

The most remarkable heavy-element assemblage on Earth is life. Plants and animals are widespread, both on the land and in the sea, though evidence of this life is not easy to detect from space. Of particular interest, the heavy-element concoctions known as men and women have existed within only the last one-tenth of one percent of Earth's history.

Life everywhere now seems biologically adapted to the planet, but adaptation is a never-ending effort. Change is inevitable. Nothing remains immutable, nothing at all. Even the rock-solid aspects of sturdy Earth transform, evolving over time scales immense compared to human life spans. What we can't see is tough to believe. But we see so briefly in time; even the duration of our civilization is a mere flicker in the spectacle of cosmic change.

Successful change is often subtle change. Indeed, large alterations in the geological, chemical, thermal, or political environment could render planet Earth uninhabitable.

∞

Imagine the conditions on Earth several billion years ago. In nearly every respect, primordial Earth and its environment differed substantially from the world we now inhabit.

Earth's original atmosphere almost assuredly contained all the most abundant elements — hydrogen, helium, nitrogen, oxygen, neon, carbon — as well as a long list of trace elements. These gases reflect those of the interstellar cloud from which our Solar System formed. This primitive atmosphere didn't stick around very long, however. Earth's surface was considerably hotter during its first billion years than it is today, and many of the atmospheric gases prevalent then must have evaporated to outer space. Gravity just couldn't hold back the early hot gases.

Despite the depletion of Earth's original atmosphere, one obviously surrounds our planet today. We wouldn't be here if it didn't. Hence, the air we now breathe must be a secondary atmosphere acquired by our planet at a later date.

For the same reason that ice cubes congeal from the outside in,

the surface of the gradually cooling primordial Earth would have been the first part of the molten planet to solidify into rock, leaving intense heat trapped below the crust. The result was surely volcanoes, geysers, earthquakes, and a variety of other geological events that literally blew off steam and pent-up heat through cracks in the surface. Outgassing of this sort happens even today, though at only a few locations on Earth, and rather infrequently at that. But several billion years ago, this type of geological activity was surely more widespread and frequent. Observations of modern volcanoes show that lots of gaseous water vapor, carbon dioxide, and nitrogen would have undoubtedly emerged, along with vast quantities of ash and dust. Smaller amounts of hydrogen, oxygen, carbon, and other gases doubtless accompanied these planetary eruptions. In this way, a new atmosphere exhaled from fissures in our restless planet.

This secondary atmosphere gradually stabilized. As early Earth cooled, the atmospheric gases were less restlessly driven by the heat that had previously caused their escape into space. Outgassed nitrogen remained in the atmosphere, where it now constitutes the largest fraction of our air. Gaseous water vapor transformed into liquid water, as Earth slowly but surely secreted its oceans. And discharged carbon dioxide reacted with silicate rocks in the presence of water to form limestone. Whatever pure oxygen gas existed on primitive Earth would have quickly vanished by reacting either with hydrogen to make more water, or with surface minerals to form oxides such as rust and sand now found throughout the crust of our planet. Breathable oxygen and the protective ozone layer arose only much later, after plants had blossomed across the face of our planet.

Shaded by Earth's secondary atmosphere, some of the outgassed chemicals would have been able to interact with one another. No coercion by outside influences was needed for the airy gases to collide, stick, and react, thus forming slightly more complex gases of ammonia and methane. Laboratory chemists verify these reactions almost every day in the course of their industrial and academic duties. And theorists well understand the electromagnetic forces among electrons that persuade these simple atmospheric atoms to combine readily, thereby concocting stable gas molecules.

With time, the molecular products of these spontaneous reactions became the interactants of additional chemical reactions. These additional reactions, however, were not spontaneous. Laboratory experiments show that the simple molecules of ammonia, methane, and water vapor require some energy in order to combine further. This energy is, in some sense, a catalyst in the production of even bigger molecules. Actually, it's more than just a catalyst. The application of energy fashions a near miracle: It synthesizes molecules a lot more complex than those likely to form by chance in a collection of free atoms and simple molecules. Astonishingly, the molecules produced are the very building blocks of life.

∞

Life may appear biologically, socially, and culturally complex, but physically and chemically it's rather simple. When divided into component parts, the basic ingredients of life — any life, from bacteria to ostriches — are no more exotic than two dozen molecules of moderate complexity. So to understand the very basic properties of life, it's unnecessary to examine an organism as complex as the entire human body; the molecular nature of contemporary life will do.

All living systems are composed of cells, the simplest form of material coagulation having the common attributes of life — birth, metabolism, and death. From the simplest organisms to the most intelligent humans, the basic unit is the cell. To appreciate chemical evolution — the changes that occurred among atoms and molecules in order to produce life — one need only consider the construction of the first cell.

Cells are small, about a million times as small as a millimeter, and thus invisible to the naked eye. About a hundred such cells would fit within the period at the end of this sentence. A microscopic view shows a central nucleus to be the most complex part of a cell. Containing trillions upon trillions of atoms and molecules, such biological nuclei should not be confused with the much smaller atomic nuclei produced directly in the hearts of stars. Resembling the yolk of an egg, the biological nucleus is surrounded by a thick, fluidic cytoplasm of less complexity. The entire unicellular life form is encased in a semipermeable membrane through which atoms and molecules can pass in and out.

Cells, then, are the simplest forms of life. However, they are vastly more complex than the simplest forms of matter — elementary particles within atoms.

The very simplest creature of all, the amoeba, consists of only one cell. More advanced organisms usually contain gargantuan clusters of cells. A grown human, for example, harbors nearly a hundred trillion microscopic cells. Their density throughout the human body averages more than a hundred million cells per cubic centimeter. Each one of these cells furthermore contains very large numbers — trillions or more — of atoms and molecules. So the density of basic matter in advanced life forms is indeed great.

Over the course of time, even as little as a second, large numbers of cells are destroyed as a result of the normal process of aging and death. All living organisms are nonetheless able to maintain a reasonably constant size and appearance throughout adulthood.

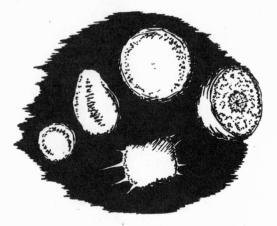

Thus, while some cells are dying, others must be forming. Our bodies, and the bodies of all other living creatures, continually manufacture more cell nuclei, cytoplasm, and membrane to sustain themselves through life. They do so by means of a curious interaction between the two basic building blocks of life.

The most important ingredients of the cytoplasm are proteins, for they comprise the foundation of any living system. The word *protein* derives from the Greek, meaning "of first importance." It's not the name of a particular substance, but rather a term for an entire class of complex biological molecules. There are tens of thousands of different proteins in the human body alone. They constitute the bulk of the guts, skin, bones, hair, and muscles, as well as every other branch of our bodies.

Proteins embrace large quantities of the element carbon. In fact, fifty percent of the dry weight of our bodies is carbon. Such "organic" substances strongly contrast with things obviously not living, such as coins, concrete slabs, or a pinch of salt. Those things are said to be "inorganic," for they are composed mostly of minerals. They contain no proteins, and their carbon content often amounts to no more than about one-tenth of one percent of their total weight.

So, carbon plays an important role in living systems. It plays a vital role in the construction of proteins.

Of what are the proteins composed? Besides their preponderance of carbon, is there a common denominator among the myriads of different proteins found throughout the wide spectrum of cells in contemporary life? The answer is yes, for experiments have shown that a rather small group of molecules make up all proteins. These basic molecules, called amino acids, are acidic (reactive) in solution and usually contain ammonia. There are only twenty of these structural units, and together they comprise all the protein in Earth's life — not just human life; all life. Amino acids are one of the two basic building blocks of life.

Amino acids are not very complex substances. The simplest, glycine, is a molecular assemblage of five atoms of hydrogen, two of carbon, two of oxygen, and one of nitrogen. Each of these atoms is held to the others by electromagnetic forces of attraction, also called chemical bonds. The most complex amino acid is tryptophan, composed of twelve hydrogens, eleven carbons, two oxygens, and two nitrogens.

In principle, the simplest possible protein should theoretically be the combination of two glycine amino acids. An electromagnetic link can be achieved, provided that a hydrogen atom is removed from one glycine, and oxygenated hydrogen from the other. This amounts to an extraction of water, and guarantees a strong chemical bond between the two glycines.

In practice, life is a little more complex, for biochemists are unaware of any real protein as simple as this. One of the simplest known proteins is insulin, with fifty-one amino acids linked together like pearls on a necklace. Another well-known protein is hemoglobin, the prime component of human blood cells. Containing nearly six hundred amino acids, hemoglobin incorporates in its structure all but one of the twenty different types of amino acids.

The biochemical function of hemoglobin (as well as all other proteins) is highly specific, and points up an important facet of protein structure. It's well known from experience in blood trans-

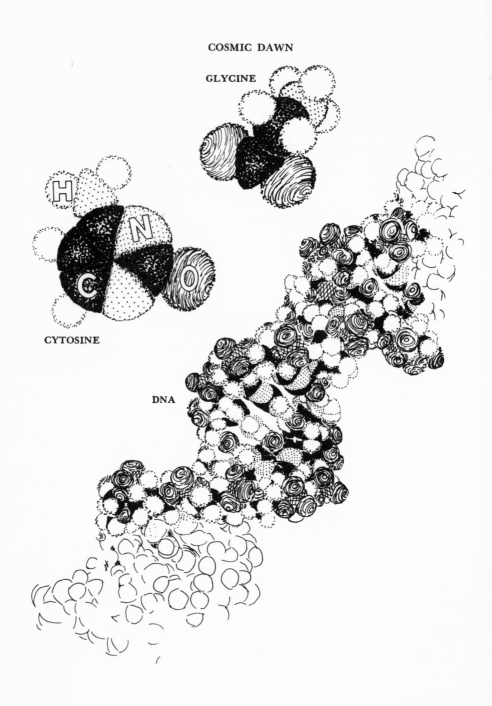

COSMIC DAWN

GLYCINE

CYTOSINE

DNA

fusions that blood of one type cannot serve as a substitute for blood of another. The differences among the various blood types result from the ordering of the amino acids. Thus, the physical and chemical behavior of a protein — just a long, stringy accumulation of amino acids — depends not only on the number of acids but also on the ordering of the acids comprising that protein.

In the larger realm, proteins give some character to cells, and cells in turn to entire living organisms. Ultimately, the fundamental character of life depends on the number and ordering of the amino acids. Only this numbering and ordering distinguishes man from mouse, demonstrating that the nature of life itself is not completely incomprehensible — at least at the microscopic level.

Mindful of the molecular structure of proteins, we return to our original concern: How is it that proteins are created in living organisms so as to maintain the integrity of those organisms? Specifically, what biochemical process serves to link amino acids so they can replenish the dead cytoplasm in all of contemporary life? Whatever the process, it must be important, since the manufacture of protein is absolutely vital to an organism's well-being — not any old protein, but exactly the right type, with its amino acids strung along in precisely the right order. To gain an appreciation for the protein construction sequence, we defer to the nucleic acids, another of life's basic ingredients.

Nucleic acids, like proteins, are long chainlike arrangements of carbon-rich molecules. Their name derives from the fact that these acids were first found in the biological nuclei of cells. Though we know of a large variety of them, the nucleic acids, again like proteins, are composed of only a small number of key constituents. Called nucleotide bases, these are the second group of life's basic building blocks. The biochemical role of the bases can be illustrated by considering the most famous of all nucleic acids — deoxyribonucleic acid, alias DNA.

DNA molecules are made of four fundamental bases — adenine, cytosine, guanine, and thymine. There is a fifth nucleotide base, named uracil, which is used in the construction of other nucleic acids, though not in DNA. These five bases play the same role for nucleic acids as do the twenty amino acids for proteins. Each nucleotide base is not much more complex than the amino acids, also being molecular assemblages of hydrogen, carbon, oxygen, and nitrogen atoms.

Although many people regard DNA as a most complex molecule, it's really nothing more than a chain of the four types of bases, which form the "rungs" of a long, thin coagulation resembling a twisted ladder. Each rung of a DNA molecule consists of two interconnected bases, giving this nucleic acid its famous double-helix structure. Experimental evidence shows, however, that all four bases do not link together equally well. Cytosine always links with guanine, forming one of the two possible base pairs, while adenine links only with thymine, fabricating the other. The structure of the molecules and especially their electromagnetic forces make incompatible any other combinations.

DNA is only one of many nucleic acids, but it stands above all the rest because of one extraordinary capability. DNA can make a copy of itself — in effect, replicate. A DNA molecule can split apart by unzipping right up the middle of the ladder. Nucleotide bases floating freely in the cell nucleus are then able to link up with each of the broken strands. The result is two DNA molecules, whereas formerly there was only one. The fact that cytosine can link only with guanine, and adenine only with thymine, ensures that the two progeny are identical to the originally split "parent" DNA molecule. The newly assembled DNA molecules then separate, and retreat to opposite sides of the cell nucleus, after which the cell divides into two, each newly fashioned cell housing a complete set of DNA molecules.

Preservation of the exact structure of the original DNA molecule is the most important feature of replication. All the information about the specific duties of that type of cell — whether a blood cell, hair cell, muscle cell, or whatever — passes from an old cell to a newly created one. Accordingly, the biological function of the "daughter" cell remains identical to that of the "parent" cell. In this way, DNA molecules, often called genes, are responsible for directing this inheritance from generation to generation.

As in the case of amino acid sequences in proteins, the ordering of nucleotide bases as well as their number is paramount in the construction of nucleic acids. The sequence of bases along a nucleic acid molecule specifies the physical and chemical behavior of that particular gene. All the genes of a living system collectively make up a genetic code — an encyclopedic compilation of the physical and chemical properties of all the cells and all their functions. In a very real sense, the structure and behavior of living organisms are dictated by the nucleic acid molecules in the nuclei of cells, for these are the material structures passed from one generation of cells to the next.

In analogy with another type of information storage — this book, for example — the individual bases might be considered words, base pairs a sentence, and an entire DNA molecule a book of instructions. The words and sentences must be in the right order to give meaning to the book. A whole library of such instructional books then constitutes the genetic code for all the varied functions performed by any living organism. In short, a full set of DNA molecules is really information — a blueprint, a master plan for every life form.

The nature of all living creatures is ultimately prescribed by the structure of the DNA molecules contained within them. The make-up of DNA molecules dictates, not only how one type of organism differs from another in both structure and personality, but also how the physical and chemical events inside a cell properly coordinate so that the activity of the cell is as it ought to be. At first glance, it would seem impossible that one molecule could do all this. But the DNA molecule is the largest known. In advanced organisms such as humans, the DNA molecule, though limited to four types of bases, can have as many as a hundred million bases or ten billion separate atoms, rendering the molecule nearly a meter (three feet) long if extended end to end. In the above analogy where a DNA base equals a word, a single DNA molecule would become the equivalent of a hundred-page manuscript. Gargantuan numbers of possible combinations of bases guarantee a vast array of diverse living creatures, each with a different appearance, style, and personality. Yet, at the microscopic level, all creatures — without exception — are composed of absolutely the same few amino acids and nucleotide bases, the very building blocks of life as we know it.

The common molecular structure pervading all life on Earth is the best evidence that every living thing arose from a single-celled ancestor billions of years ago.

Continuous production of the cytoplasm's proteins makes heavy use of the cell's nucleic acids. The sequence of events goes roughly as follows: Just prior to cell division, the DNA molecule dispatches a relative molecule, called RNA, from the biological nucleus to the cytoplasm. RNA stands for ribonucleic acid — a smaller, single-stranded version of the normally double-stranded DNA. The RNA molecule acts like a messenger, carrying instructions from the DNA molecule. Once in the cytoplasm, single-stranded RNA attracts freely floating amino acids to its uncoupled bases. Only certain amino acids can successfully attach to the RNA bases, for the electromagnetic forces of those bases attract some amino acids, while repelling others. After a certain amount of time — on the order of microseconds — the single-stranded RNA molecule secures, by accidental collisions, an entire complement of amino acids. These acids also attach to one another by means of their own electromagnetic forces. Finally, when the long chain of amino acids is fully assembled along the entire length of the messenger RNA molecule, the chain detaches and drifts off into the cytoplasm of the cell. A protein has thus formed. But this is no ordinary protein created at random. Rather, it's a specific protein constructed according to the instructions provided by the RNA molecule. In this way, RNA acts as a prescription, or a template on which protein molecules are built — a template originating in the cell nucleus with the DNA molecule itself.

This, then, is the way in which proteins are continually replenished in living organisms. All living systems grow and eventually

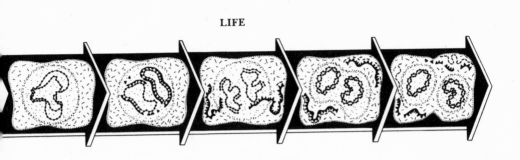

become biologically stabilized in this same way. The whole process is occurring repeatedly in our bodies right now. Of course, different organisms have different DNA structures and therefore manufacture different types of protein. In fact, no two living organisms have the same DNA structure, except for identical twins.

Whether in man, mouse, or daisy, the nucleic acids mastermind life, and the proteins maintain its well-being.

∞

Understanding contemporary life is one thing, but appreciating how it might have arisen from nonliving matter is quite another. Can we be sure that the basic ingredients for life were present on primordial Earth? Furthermore, can we prove that those nonliving building blocks could have fashioned a simple living cell given the harsh conditions on our planet billions of years ago? These questions can only be addressed in the laboratory, for the atmosphere and surface of today's Earth differ vastly from those of early Earth. Results of modern laboratory experiments that mimic the geophysical environment on primordial Earth imply that the answer to these questions is yes.

Utilizing a large test-tube-like contraption holding water and various gases, a rather simple laboratory experiment can be undertaken to simulate Earth's early ocean and atmosphere. The gas — usually a mixture of ammonia, methane, hydrogen, and sometimes carbon dioxide — is meant to reproduce the composition of the secondary atmosphere. Though toxic to present-day life, some blend of this gas was apparently just right for the origin of life.

Likewise, the liquid is meant to resemble the primordial oceans or some such pool of water. Upon heating of this "ocean," water vapor rises to mix with the other gases in the "atmosphere," whereupon it eventually "rains" back down with any newly formed chemicals. The apparatus is shut tight, allowing all the gases to cycle endlessly without escaping, much like the familiar evaporation-condensation sequence happening every day on Earth. In the absence of energy, the gases do just that — cycle through the machine unchanged. These gases refuse to react spontaneously with one another. For example, when molecules of methane and water vapor make contact, nothing happens. They need a little help in order to react chemically. And that help, that catalyst, is energy of some sort. Once energy is applied, some of the bonds within each of the gas molecules break apart, thereby allowing the liberated atoms and molecular fragments to reform as more complex molecules.

Sometimes, in order to hasten the reactions, researchers employ greater gas abundances than are thought to have existed on Earth long ago. Or they increase the intensity of the energy source above the amount presumed present billions of years ago. As a result, the molecules' chance of colliding with one another is enhanced enormously, allowing the experimental simulations to be completed in a few weeks. Quite frankly, researchers with finite careers and one-year grants can't afford to wait several hundred million years to determine the outcome of their experiments.

After a few days of energizing the gases, a thick, reddish brown, soupy material appears in the trap at the bottom of the apparatus. Chemical analysis shows this slimy gunk to contain molecules considerably more complex than those initially inserted into the apparatus. Be assured, nothing living crawls out of this primordial soup. But several of the molecular products are among the known precursors of life. They include many of the amino acids and nucleotide bases comprising the building blocks of all contemporary life.

The recipe for the successful creation of the prelife acids and bases is not a very rigid one. You could probably undertake the experiment in your bathtub, though it's not recommended. The

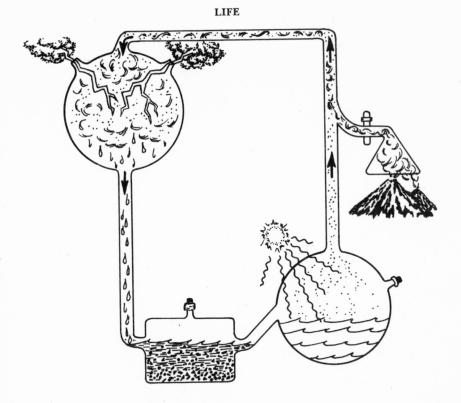

simple gases, energy sources, and cooking times have been widely varied by different investigators throughout the past two decades. Provided there's no free oxygen gas present, the result is invariably the synthesis of complex organic molecules. With even small doses of oxygen in the test tube, however, the gases oxidize, the concoction becomes unstable, and none of life's ingredients takes shape. Ironically, although most of Earth's established life today requires oxygen, this gas would have acted like a deadly poison during the formative stages of that very same life.

A critical consideration here concerns the amount and kind of energy used to power these experimental simulations. Is it reasonable to suppose that enough of the right type of energy was present on early Earth? Usually the laboratory simulations employ energy provided by electrodes sparking the gases in the test tube. Electrode flashes can be thought of as imitation lightning. This spark discharge can also be regarded as an indirect simulation of several other types of energy undoubtedly present on early Earth. Besides atmospheric lightning, there was surely plenty of radioactivity and volcanic activity, both of which produce heat. Even thunder guarantees enough energy to have powered in Earth's early atmosphere some of the chemical reactions known to occur in the laboratory experiments; if thunder can break windows, it can also break chemical bonds. Meteoritic bombardment is a further source of energy; as huge rocks plow through the atmosphere, their shock waves generate heat often sufficient to ignite chemical reactions.

Most of these energy sources are localized, and hence were sufficiently intense to make or break molecular bonds only at isolated locations on early Earth. Solar energy, however, was widespread. While ordinary sunlight is not energetic enough to trigger chemical reactions, solar ultraviolet radiation is. Without oxygen, an ozone layer could not have surrounded prelife Earth, and this ultraviolet radiation would have had little trouble reaching Earth's surface. It would seem that solar energy, which clearly sustains life now, was also the dominant nutrient in the creation of life billions of years ago.

Laboratory experiments like these are significant because they demonstrate that the molecular building blocks of life could have been synthesized in any one of many different ways during the early history of planet Earth.

These basic ingredients, however, are not life itself. They are still much simpler than a single cell. Amino acids and nucleotide bases are in fact substantially less complex than even the proteins and nucleic acids found throughout contemporary life. How, then, were the ingredients of this primordial soup initially assembled into proteins and nucleic acids? Apparently, the soupy organic slime was further concentrated so as to permit not only stronger but also drier interaction.

As noted earlier, two amino acids can be linked to reach the next stage of complexity, provided a water molecule is removed. Such a dehydrated condensation of many amino acids can then construct chain molecules into complex proteins. Successive linkages of nucleotide bases can similarly fashion lengthy nucleic acids.

Heat, for example, could have evaporated some water, especially along the shoreline of a primordial ocean or a lagoon inlet. It seems reasonable to suppose that the repeated in-and-out sloshing of tides in shallow waters could have led to a daily cycle of solar dehydration of molecules in a temporarily dried lagoon during low tide, followed by further interaction of those molecules in the open ocean during high tide.

The opposite condition — cold — can also effectively remove water molecules from an organic mixture. The freezing of water transforms it from liquid to ice, thus allowing the acids and bases to become more concentrated and hence linked together. Repeated freezing and thawing could allow the buildup of progressively larger chain molecules.

A third mechanism can actually remove water while still in the presence of water. Although this sounds impossible, it's done all the time in living organisms; composed largely of liquid, the cells in our bodies routinely manufacture protein. They do it by using catalysts, third-party molecules that act like brokers. Catalysts that speed condensation reactions in contemporary life were probably absent in the primordial ocean. Yet many investigators speculate that other catalysts would have existed four billion years ago to promote such condensation reactions in water. Certain types of clay, for example, are thought by many researchers to have been the scaffolding required to fabricate larger organic molecules along the edges of oceans, lakes, and rivers.

Scientists are unsure that the first complex proteins and nucleic acids really did originate in any of these ways. It's unlikely that the fossil record will ever show the precise path whereby prelife molecules gradually coagulated into something that might be genuinely called life. Heating, freezing, and catalyzing are nonetheless plausible agents for the self-construction of amino acids and nucleotide bases into proteins and nucleic acids.

* * *

The simple cell remains astonishingly more complex than any of these prelife molecules. To reach this very root of the evolutionary tree, biochemists currently seek a detailed understanding of how the proteins and nucleic acids were in turn able to forge more intricate combinations of biological significance. Understanding in this area is limited, however. Scientists have only been able to surmise that repeated interactions among the many molecules on primordial Earth could have eventually produced something resembling today's proteins, DNA, and simple cells.

More advanced laboratory simulations of recent years support this view. Repeated energizing and dehydrating under the simulated conditions of primordial Earth produce organic molecules more complex than the amino acids and nucleotide bases. In particular, dense clusters of amino acids, called proteinoid microspheres, can be observed through a microscope. Only about a hundredth of a millimeter across, these clusters are not well-known proteins such as insulin or hemoglobin, but protein-resembling compounds whose relevance to the origin of life is uncertain. Direct chemical analyses confirm that these microspheres are indeed dense coagulations of organic material floating in a mostly inorganic fluid. A microscopic view shows them shimmering like globs of oil on the surface of water, or grease on the surface of cooled chicken broth. Some researchers regard proteinoid microspheres as bona fide proteins, while others are not so sure.

In many remarkable ways, the proteinoid microspheres produced in laboratory experiments resemble simple bacterial cells. They seem to possess a semipermeable membrane through which small molecules can enter from the outside in order to "feed" the microsphere, but which is not penetrable by most larger molecules synthesized within. Some discharge of wastes can be observed through the microscope but, by and large, there's a net intake of material, in some ways mimicking contemporary biochemical cells. Indeed, these curious little bags of chemicals could be regarded as capable of eating, growing, and excreting — much like a primitive metabolism, perhaps.

When the experimental apparatus is jostled to create some turbulence in the fluid — the analogue of early oceanic wave action — the larger proteinoid microspheres are observed to break into smaller ones, demonstrating what some workers regard as a primitive form of replication. Some microspheres disperse, an apparent "death." Others grow like their "parents," only to be ruptured by another act of "replication."

So the proteinoid microspheres have a primitive resemblance to simple bacterial cells. Some eat, some grow, some reproduce, and some die. Can these microspheres be called life? It's hard to say. The distinction between matter and life is not clear-cut; it's fuzzy because life itself is nearly impossible to define.

Most biologists argue that amoebas are definitely living, but that the molecular contents of the organic soup are not. Proteinoid microspheres apparently lie somewhere in between. But if they are not at least progenitors of Earth's living systems — protolife — then nature would seem to have played a malicious trick on modern science.

∞

All these theoretical considerations and experimental simulations suggest that life is a logical consequence of known chemical principles operating within the atomic and molecular realm. Furthermore, and more fundamentally, the origin of life itself seems to be a natural consequence of the evolution of those atoms and molecules.

These statements thus far lack absolute proof, since laboratory experiments have yet to fashion anything more sophisticated than life's precursors. To be sure, a large gap separates these preliminary stages of chemical evolution and the onset of biological evolution of simple living cells. Biochemists have nonetheless formed a consensus concerning that blurred realm where chemical evolution ends and biological evolution begins.

Energy is the one thing required in any aspect of evolution, regardless of whether that evolution involves matter that is clearly

165

living or matter that is clearly not. Neither matter nor life can proceed from a simple to a complex state without absorbing some energy. Complex objects have some organization, and organization of any kind requires energy. Even when fully formed and highly evolved, no advanced matter, whether in stars or people, can sustain itself without a continual supply of energy. This energy is a fuel, a food of sorts.

In the case of the laboratory simulations just described, energy derived from the spark discharge can be considered an "explosive food" used to fracture bonds of the small molecules, thereby enabling them to reunite as more complex bunches of atoms. Part of the spark's energy is also absorbed. Much of it strengthens the chemical bonds necessary to hold together — to organize — the new, more complex acids and bases. The organic scum floating on the surface of the primordial ocean thus became a tremendous storehouse of energy.

Repeated energizing — that is, repeated feeding — is required for the construction of proteinoid microspheres. Once formed, these organic droplets require even additional feeding to maintain their increasingly intricate molecular organization. They feed on nutritious amino acids and nucleotide bases admitted through their semipermeable membranes; the microspheres extract energy by chemically breaking bonds among the atoms comprising those acids and bases. In this way, proteinoid microspheres essentially "eat" by absorbing minute amounts of energy from their surroundings.

Why do the proteinoid microspheres obtain energy from their immediate environment? Why don't they continue to utilize one of the external forms of energy, such as solar radiation, atmospheric lightning, or volcanic activity? The answer is that the energy used to create microspheres is often too harsh to sustain them. As molecules become larger and more complex, they also become more fragile. They eat and organize by absorbing energy, but that energy must be slight and gentle. (It's a little like the difference between watering a plant and drowning it.) The small acids and bases able to pass through the miniature openings of a microsphere's membrane contain just the right amount of energy. They enable proteinoid microspheres to survive without being subjected to the harsh external energy required for their earlier synthesis.

Although scientists have no direct evidence for the assembly of more advanced precursors of life, laboratory studies strongly support a two-step process like that outlined above: a heavy dose of energy was required to fashion the precursors, after which milder energy was needed to maintain them.

Circumstantial evidence and biochemical insight lead scientists to surmise that proteinoid microspheres, once formed, were able to secure protection from the uncontrolled energetic conditions that created them. This is not unreasonable, since Earth was rapidly cooling several billion years ago, becoming in the process geologically more quiescent. As time passed, volcanoes, earthquakes, and atmospheric storms would have gradually subsided. The amount of solar ultraviolet radiation reaching the ground would also have diminished as terrestrial outgassing thickened the atmosphere. Many prebiological substances probably found shelter under thin layers of water capable of absorbing whatever harsh solar radiation did manage to penetrate the air.

From this point on, biologists can only presume that at least one proteinoidlike amalgam was eventually able to evolve into something everyone would agree is a genuine living cell. Be sure to recognize, though, that nothing yet discovered in the fossil record documents this prelife evolutionary phase. Furthermore, substances more complex than the proteinoid microspheres have not been synthesized in laboratory simulations of primordial Earth conditions; such microspheres possess neither the hereditary DNA molecule nor a well-defined nucleus common to most contemporary cells. Alas, researchers cannot presently explain how the first protein might have arisen from a medium containing no nucleic acids, especially when the passage of information from nucleic acid to protein is considered by many to be the central dogma of modern molecular biology.

The question of which came first, proteins or nucleic acids — that is, "protobiont" or "naked gene" — resembles the familiar chicken-or-the-egg paradox, and clearly represents one of the biggest bottlenecks in all of cosmic evolution. Presumably, the capacities for metabolism and reproduction developed in parallel, but we don't know for sure. Frankly, there's a considerable gap in our direct knowledge of the precise events that occurred between the synthesis of life's precursors and the appearance of the first bona fide cell.

The fuzzy interval between life and nonlife troubles scientific researchers and laypersons alike. The central tenet of chemical evolution is straightforward enough: Life has evolved from nonlife. But aside from biochemical intuition, and laboratory simulations of some key events on primordial Earth, is there anything at all in our inventory of knowledge suggesting that life may indeed have developed from nonliving molecules? Fortunately, the answer is yes.

The smallest and simplest entity that sometimes appears to be alive is called a virus. We say "sometimes" because viruses seem to possess the attributes of both nonliving molecules and living cells. Named after a Latin word for "poison," viruses are regarded as the cause of disease. Though they come in many sizes and shapes, all viruses are smaller than the typical size of a modern cell. Some are in fact made of only a few hundred atoms. At least in terms of size, then, viruses seem to bridge the gap between cells that are living and molecules that are not.

Viruses contain both protein and DNA, though not much else — no unattached amino acids or nucleotide bases by means of which living organisms normally grow and reproduce. How then can a virus be considered alive? Well, it's not when alone; a virus is absolutely lifeless when isolated from living organisms. But when in contact with living systems, a virus has all the properties of life. It comes alive by injecting its DNA into the cells of normal living organisms. The genes of a virus seize control of a cell, and establish themselves as the new master of chemical activity. Viruses grow and reproduce copies of themselves by using the free acids and bases of the invaded cell, thereby robbing the cell of its usual function.

Rapidly and wildly, viruses multiply, spreading disease and, if unchecked, eventually killing the invaded organism.

Researchers are thus unable to classify viruses as either living or nonliving. Even in the contemporary world, life seems to shade almost imperceptibly into nonlife. Viruses apparently exist within this shaded, uncertain realm.

What were the first living cells like? Scientists don't know for sure. Most likely they were tentative microscopic entities — fragile enough to be destroyed by strong bursts of energy, yet sturdy enough to reproduce, thereby giving rise to generations of descendants.

One thing is certain: The first cells, called heterotrophs, somehow had to find enough energy to continue living and organizing themselves. They must have done so while floating on or near the ocean surface, absorbing the acid and base molecules of the rich broth of the organic ocean. This extraction of energy via the capture and chemical breakdown of small molecules is known as fermentation, a process still employed by bacteria (in wine casks, bread dough, and elsewhere) on Earth today. But the heterotrophs could not have indefinitely scavenged enough of the organic materials from which they originated. Why not? Because the passage of time brought irreversible changes in the environment.

As Earth's heat diminished, so did several of the energy sources capable of producing acids and bases. Whereas originally there had been plenty of juicy organic molecules on which the heterotrophs could feed, the thickened atmosphere and decreased geological activity ensured fewer food supplies. Consumed more rapidly than it was replenished, the organic soup gradually thinned, so much so that the primitive cells had to compete with one another while scrounging for the diminishing supply of nourishing acids and bases. Eventually, the heterotrophs devoured every bit of organic material floating in the ocean. The organic synthesis of acids and bases via sunlight, lightning, or volcanoes simply couldn't satisfy the voracious appetites of the growing population of heterotrophs.

This scarcity of molecular food was a near-fatal flaw in life's early development. It was an ecological crisis of the first order. Had nothing changed, Earth's simplest life forms would have proceeded toward an evolutionary dead end — starvation. Earth would be a barren, lifeless rock. Fortunately, something did change. It had to; nothing fails to change.

Some primitive cells — the forerunners of plants — were granted a new way to get energy, thus conceiving a unique opportunity for living. This novel biological technique employed carbon dioxide, the major waste product of the fermentation process. While the earliest cells were busy eating organic molecules in the sea and as a result polluting the atmosphere, more advanced cells were able to use these pollutants to extract energy. In this case, the energy was not derived from the carbon dioxide gas, but from another well-known source — the Sun.

The key here is the molecule chlorophyll, a green pigment having its atoms arranged in such a way that light, striking the surface of a plant, is captured within the molecule. Advanced cells containing chlorophyll can extract energy from ordinary, gentle sunlight (not harsh ultraviolet radiation) by the process of photosynthesis. This is a chemical reaction that exploits sunlight to convert carbon dioxide and water into oxygen and carbohydrates. The oxygen gas escapes into the atmosphere, while the carbohydrate (sugar) is used for food. This, then, is another way a cell can "eat," or extract energy from its environment.

How did some protoplant cells come to develop photosynthesis? Scientists again don't know for sure, other than to suggest that chance mutations altered some earlier cells' DNA structure. Those whose DNA molecules allowed photosynthesis were able to survive on solar energy. They didn't need to compete for the diminishing supply of oceanic acids and bases. In short, these so-called autotrophs were able to adapt to the changing environment. They had an advantage, since they were clearly more fitted for survival during what was probably the first ecological emergency on our planet.

Photosynthesis freed the early life forms from total dependence on the diminishing supply of organic molecules in the oceanic broth. Early cells able to utilize sunlight overspread the watery

Earth. In time, the autotrophs evolved to become all the varied plants now strewn across the face of our planet.

The photosynthetic reaction has, in fact, survived to this day as all plants routinely use sunlight to synthesize carbohydrates as food, and in the process release oxygen gas. Photosynthesis is the most frequent chemical reaction on Earth. Each day about four hundred million tons (about a trillion kilograms) of carbon dioxide mix with some two hundred million tons of water to produce about three hundred million tons of organic matter and another three hundred million tons of oxygen.

By loss of their food source, the ancient and primitive heterotrophs were naturally destined to die. The autotrophs were naturally favored to live. Life on Earth was on its way toward using the primary and plentiful source of energy — solar radiation. It all began some three billion years ago.

The operation of photosynthesis over eons is, by the way, partly responsible for the fossil fuels. Dead, rotted plants, buried and squeezed below layers of dirt and rock, have chemically changed over megacenturies into oil, coal, and natural gas. Such fossil fuels, with vast quantities of energy intact, have made industrial civilization possible. But they are virtually nonrenewable, at least over time scales shorter than tens of millions of years. Indeed, billions of years of energy deposits in dead plants will be depleted shortly. Once again, things will have to change, just as they've changed in the past.

The use of sunlight by cells was a double achievement of great importance for life on Earth. Not only did it provide an unlimited source of energy and assure a dependable supply of food, but it also drastically altered Earth's atmosphere by endowing it with lots of oxygen gas.

Atmospheric oxygen has had an enormous influence on the abundance and diversity of life on Earth. Why? Because the photosynthetic release of oxygen molecules into an atmosphere that previously had little or none of it ensured great changes in the environment. Upon interaction with the Sun's ultraviolet radiation, the diatomic oxygen molecule breaks down into two oxygen atoms. Three oxygen atoms then combine near the top of the atmosphere, molding large quantities of triatomic oxygen called ozone. (*Ozone* is derived from the Greek and means "to smell"; the pungent ozone gas can often be sensed near thermal copying machines that use ultraviolet radiation.) Ozone now completely surrounds our planet in a thin shell at about fifty kilometers (thirty miles) altitude, effectively shielding the surface from further exposure to harmful ultraviolet radiation.

As the ozone layer matured, survival no longer depended on protection by a layer of water, or by some rock or other object acting as a barrier against the harsh realities of the real world. Life was then free to populate all the available nooks and crannies on planet Earth. In short, life could invade areas where no life had existed before.

None of this happened overnight. The ozone screen needed time to become established. The whole process was an accelerating one: Oxygen-producing autotrophs had an increased chance for survival and therefore replication. The more offspring they produced, the more oxygen was dumped into the atmosphere. And more oxygen meant more ozone. But it still took time for the protective ozone to develop. How long? Perhaps several billion years after the onset of photosynthesis.

Some calculations suggest that the ozone layer didn't become a really efficient shield of solar ultraviolet radiation until six hundred million years ago. There's also solid evidence that life suddenly became varied and widespread around that time. Prior to six hundred million years ago, only primitive life forms existed. Time periods shortly thereafter saw a rapid surge in the total number and diversity of complex living organisms — a population explosion of the first magnitude.

What was responsible for this burst of biological activity? It was largely the presence of oxygen, which permitted a new, more efficient way for organisms to obtain energy for living. The first, most primitive life forms that ate via fermentation were superseded by more advanced life forms that developed photosynthesis as a means to manufacture food. Eventually, more sophisticated organisms — the forerunners of animals — began exploiting oxygen as their primary source of nourishment. By using oxygen, organisms could obtain more energy from the same amount of food. Combined with a fully established ozone shell, this global availability of oxygen meant that life was then able to survive and reproduce in all sorts of new habitats.

The previously harsh conditions under which early life struggled had disappeared. Earth became a nice place in which to live. And the vanguard organisms of the time took advantage of their friendlier environment.

Respiration is the chemical process whereby cells employ oxygen to release energy. Ingesting oxygen ("breathing") helps an organism to digest the carbohydrates in its body, the waste products being carbon dioxide and water. Respiration, then, is just the reverse of photosynthesis, but there's an important difference. Whereas in photosynthesis energy must be absorbed to yield the foodstuff carbohydrates, in respiration much of that same energy is released as the oxygen decomposes the chemical bonds of those carbohydrates.

Today, these two processes — plant photosynthesis and animal respiration — direct the flow of energy and raw materials throughout the biosphere. This energy for life is unidirectional. It originates with the Sun, is absorbed in photosynthesis, is released by respiration, and is consumed in the process of living. All the while, carbon dioxide, water, and oxygen are endlessly exchanged between photosynthesis and respiration. These materials are used repeatedly, in a completely cyclical fashion; the plants use animal pollution, while the animals use plant pollution. Nature knows how to recycle.

As far as we know, the establishment of this most efficient energy flow — from the Sun to advanced organisms — contributed mightily to the veritable explosion of biological organisms several hun-

dred million years ago. Solar energy exchange endures to this day. It now energizes humans; if we're wise, it will someday energize our technological civilization.

∞

By what means do scientists know so much about the previous episodes of life on Earth? Never mind the precise details; how can we sketch even a broad outline of such ancient events? The answer is that we are guided by many clues preserved in the old rocks of our planet. Most of these hints are the remains of living organisms.

Life forms usually begin to decay as soon as they die. Once a means of gathering energy has terminated, disorder sets in. Entropy inevitably increases, according to the basic laws of thermodynamics. Dead organisms — even their bones — decompose quickly, the former proteins becoming bad-smelling substances within a few days. This is the way in which all life forms — including humans — return to the planet the elements borrowed from it.

But some special environments can limit decay, including polar regions, high mountaintops, and deep ocean bottoms. Low temperatures and water burial serve to retard spoilage. For example, a living system, having perished along a stream or ocean shore, might be buried under layers of sand and sediment as they settle down through the water. Volcanic lava is another material in which various life forms can be buried, in this case under mounds of ash. In time, the sedimentary deposits of sand or lava become hardened into rock, entombing the remains of living systems. Thus, though the fleshy parts of ancient organisms usually disappear, their bony structure is occasionally preserved until they are later uncovered by natural causes (changes in Earth's crust) or man-made events (archeological expeditions). These rare remains are called fossils, the visible traces of dead organisms that once lived.

Exhaustive study of the fossils thus far uncovered has enabled biologists to assemble a reasonably complete record. With a variety of dating techniques, they can roughly determine when various organisms lived. More specifically, the fossil record shows that new life forms periodically emerged, while others perished. Some types of life survived for long periods of time; others seem to have succumbed as soon as they appeared.

A rule of thumb suggests that the oldest rocks have only simple life embedded in them, whereas the youngest rocks contain mostly complex life. Apparently, many forms of life were able to survive only by increasing their complexity.

The oldest fossils seem to have a cellular structure resembling that of modern blue-green algae — fuzzy bacterial moss found at the edge of lakes, streams, and even backyard swimming pools. Not terribly complex, these fossils lack well-developed biological nuclei. Yet, these life forms, of which the fossils are the remains, must have photosynthesized by some means or another, since chlorophyll products are often found in their immediate vicinity. Apparently these oldest fossils are the remains of autotrophs.

Fossil remnants of the most primitive cells were discovered only within the past decade. Found embedded in African and Australian sedimentary rock known to be nearly four billion years old, these oldest cells are presumed to have existed that long ago. In truth, there's no way of dating the fossils themselves; radioactive techniques are useless for carbon substances older than forty thousand years. But it seems inconceivable that latter-day algae could have gotten so deeply encased in such old rocks.

There are no known fossils of the most ancient heterotrophs. Small molecules can seep into all but the densest of rock formations, so that there is no good way of knowing if amino acids and nucleotide bases contaminating old rocks are as ancient as the rocks themselves. Even if there were techniques enabling scientists to search for prebiotic organic matter, probably none would be found. Heterotrophs probably devoured every bit of available organic matter, leaving absolutely no trace of the primordial soup anywhere on Earth. Consequently, scientists are unable to estimate either the amount of time needed for the autotrophs to have overwhelmed the primitive heterotrophs, or for the primitive heterotrophs to have appeared in the first place. All that we can say with certainty is that life must have originated not more than a billion years after the formation of planet Earth. It conceivably could have taken less time, but how much less can only be a guess.

Evidence for more recent, though still ancient, life has been unearthed in numerous localities on our planet. The north side of Lake Superior in Ontario is especially rich in ancient fossils, and the rock formation there has been radioactively dated to be a little more than two billion years old. Embedded within the limestone are fossils called stromatolites, layered colonies of algae created when primitive autotrophs clustered together and became trapped in sediment that later hardened into rock. Careful examination of this very old Canadian rock shows evidence for at least a dozen distinctly different types of algae, all extremely simple systems compared to the complexities of today's cells. As far as we know, these two-billion-year-old cells still lacked well-developed biological nuclei. And, despite the clustering, each ancient cell presumably still functioned on its own, not in collaboration with other cells nearby. These were still unicellular life forms.

One-billion-year-old rock formations scattered across the globe often contain surprisingly well-preserved remains of autotrophs. Microfossils of many different types have been identified, several similar in structure to modern blue-green algae. In addition, fossils of this period record the appearance of the first true organisms — organized teams of cells, the ancestors of modern plants and animals.

Unicellular forms of life had, by this time, already existed on Earth for nearly three billion years. They had become larger, more sophisticated, and perhaps functionally diverse. And they had developed full-fledged biological nuclei, including hereditary DNA molecules. Most important, they were beginning to enhance their survivability by working together as multicellular units. A billion years ago, life had reached a whole new plateau — it had become organized.

But that's all there was a billion years ago. Primitive oceanic life flourished, though nothing else did. The fossils show no evidence that plants yet adorned Earth's landscape. No animals were crawling, swimming, or flying near the surface. And certainly, there were as yet no men and women.

∞

To grasp the full spectacle of life, past and present, we must probe beyond biochemical analyses of microscopic matter. Study of the single cell is sufficient to display the operational difference between life and nonlife, but it's not enough to illuminate the total spectrum of life on Earth. To decipher life's complexity, it's necessary to study entire organisms. For the same reason that no one could hope to understand the inner workings of an automobile by grinding it up and chemically measuring its constituent atoms and molecules, the macroscopic analysis of whole living things usefully complements the microscopic view of their cells.

Today, biologists recognize at least two million different kinds of plant and animal life thriving on our planet. This number encompasses the entire range of current Earth life, from tiny microorganisms to giant whales and redwood trees. New life forms are constantly being discovered. Yet even with this enormous variety of extant life, the fossil record suggests that about ninety-nine percent of all life forms that have ever resided on Earth are now extinct.

The fossil record narrates the following history of relatively recent life on planet Earth: Less than a billion years ago, once multicellular organisms had learned to utilize oxygen, they quickly evolved into highly specialized creatures. These oxygen-breathing animals, the earliest ancestors of humans, swarmed in the sea, feeding upon plants and upon one another. Some could only float on the water, others anchored themselves to undersea slopes, while still others had some mobility in the water. Almost all these creatures, alive between one-half and one billion years ago, had soft bodies. Hence, the fossil record of the earliest respiratory organisms is understandably sketchy, for without bones or shells, little of them has remained intact to this day.

More than half of all fossils are trilobites, lobsterlike creatures of enormous variety. Some had heads, some apparently not; others had a dozen eyes, still others none at all. Most were quite small, measuring a few centimeters in length, but some stretched half a meter from head to tail. Though all trilobites are now extinct, paleobiologists are reasonably sure that some version of them gave rise to all of today's animals.

As the years wore on, life-styles multiplied rapidly. Each type of organism responded to changes in the oceanic, continental, and atmospheric environment. Each attempted to adapt for better survivability. By some five hundred million years ago, worms, clams, and snails ruled the world.

The sudden richness of the fossil record about six hundred mil-

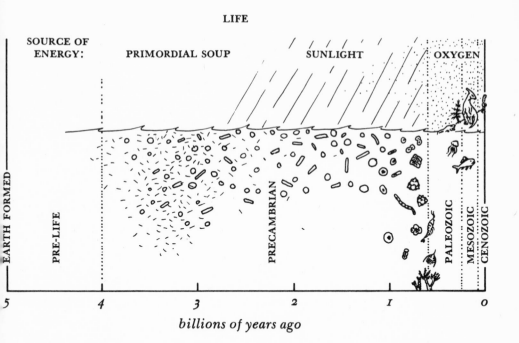

SOURCE OF ENERGY:

PRIMORDIAL SOUP   SUNLIGHT   OXYGEN

EARTH FORMED

PRE-LIFE

PRECAMBRIAN

PALEOZOIC

MESOZOIC

CENOZOIC

5   4   3   2   1   0

billions of years ago

lion years ago reflects, not only the probable establishment of the ozone layer and the photosynthesis-respiration cycle, but also the development of primitive skeletons. Paleontologists, the fossil experts, accordingly consider this to have been the start of a new age in the history of life on our planet. All Earth time longer ago than six hundred million years is called simply the Precambrian. Since then, there have been three additional ages, broadly subdividing more recent times: Paleozoic, the Greek word for "ancient life"; Mesozoic, designating "middle life"; and Cenozoic, meaning "recent life." The starting times of each are roughly 570 million, 225 million, and 70 million years ago.

The Paleozoic fossil record shows that the first fish materialized at least four hundred million years ago, the first amphibians and forests some three hundred million years ago, the first reptiles and insects shortly thereafter. Over the course of a hundred million years or so, life had spread from its oceanic nursery. It had come ashore. The inaugural exploration of the alien land was probably undertaken by sea plants that migrated along the previously barren and rock-strewn coast. Some types of fish, which depended on this vegetation, apparently followed their source of food. Some may have invaded the land intentionally, whereas others may have been washed up onto shorelines by storms or left high and dry when shallow ponds evaporated. Those species of fish that successfully negotiated the land became four-legged, air-breathing amphibians; those that tried and failed, became extinct. Details are sketchy because geologists have no record of where shorelines were so long ago; all of this preceded the breakup of the ancestral landmass that gave rise to the modern continents as we know them. The fossil record does dictate one thing for certain: Descendants of these first shore plants became the world's first forests, and certain descendants of the amphibians eventually became the animals that lived in those forests.

By the end of the Paleozoic age, life was firmly implanted in the sea, on the land, and in the air. By two hundred million years ago, there existed a broad opportunity for living. The land in particular, with its green expanses and virgin forests, enabled animal life to proliferate with astonishing diversity. Species multiplied like crazy, so much so that the fossil record documents, for example, that there were at that time already a thousand different kinds of roaches; the household version — the common cockroach — is a direct, and very durable, survivor of the late Paleozoic age. All life on Earth had become dominated by the reptiles, a whole new life form that had, over millions of years, evolved from vertebrate amphibians. The conquest of the land was complete, as reptiles spread out to fill every conceivable niche on the planet. Ancestors of nearly every animal now on Earth, the reptiles of two hundred million years ago had developed supple backbones, mobile legs, and keener brains than any other creature inhabiting Earth until that time.

The Mesozoic fossil record shows that many forms of life not only thrived but also evolved toward greater complexity. Plant life flourished, taking its final steps toward its modern forms. Flowers appeared in dazzling colors and rich scents, all for the purpose of attracting pollinating insects. And the first birds took flight, most as small as today's sparrows.

But the highlight of the Mesozoic age was the first appearance of the mammals, warm-blooded animals able to derive body heat from digested food and thus stay comfortable in cold environments. Fossil evidence reveals that three types of mammals evolved throughout this hundred-fifty-million-year period. The earliest mammals were probably ancestors of the present-day anteater — primitive creatures that had fur and nursed their young with milk, but, like reptiles, laid eggs instead of bearing live young. A second, more advanced group of mammals probably bore their young live like their descendants, the modern kangaroo and the koala bear; these young were so small and immature, however, that they had to be incubated in a fur-lined pouch under their mother's belly. Toward the end of the Mesozoic, true mammals appeared, laying no eggs and needing no pouch for their young.

This outline aside, details of the mammals' line of ascent during the Mesozoic age are somewhat obscure, as they were completely overshadowed by the mightiest reptiles of all time — the dinosaurs. The name of these monstrous beasts derives from the Greek words *deinos* (terrible) and *sauros* (lizard). Nothing at all like the snakes, lizards, or crocodiles of present times, in their prime, roughly a hundred million years ago, the dinosaurs roamed Earth with skill and power, overrunning the air, land, and sea until they completely and devastatingly dominated our planet. Dinosaurs were mostly department-store-sized land creatures, though their relatives included awesome seagoing reptiles capable of swallowing today's great white shark in one gulp, and fearsome airborne brutes having wingspans comparable to those of modern fighter aircraft. Their fossils have been uncovered on all the world's continents, except at the poles.

Until recently, the prevailing view claimed that dinosaurs were rather dumb — cold-blooded and small-brained. In chilly climates, or even at night, the metabolisms of these huge reptiles would have become sluggish, making it difficult for them to move around, secure food, and thus survive. However, a different though con-

troversial view is now emerging from many paleontological laboratories. Recent studies of dinosaur fossils suggest that these
monsters might have had large, four-chambered hearts, like those
of mammals and birds. Such a heart could have pumped blood
through organs, enabling dinosaurs to sustain a high level of physical activity. If these recent interpretations are correct, dinosaurs
were probably warm-blooded, and thus relatively fast-moving
creatures. Also, though the dinosaurs clearly had small brains
compared to those of today's mammals, they were still smart for
their time. Indeed, no species able to rule Earth for nearly a hundred million years could have been too dumb. By comparison,
humans have thus far dominated for little more than two million
years.

All these flying, swimming, or landlocked prehistoric predators
disappeared with bewildering abruptness near the end of the Mesozoic age. The presence of the dinosaurs mysteriously vanishes from
the fossil record. No one knows why. Nor does anyone know how
the mousy mammals were able to survive throughout the hundred-
million-year reign of terror brought on by these enormous beasts.
Whatever the reason, it affected not only the dinosaurs but also
many other forms of life. Fossil records demonstrate that, almost
a hundred million years ago, nearly half of all plants ceased
to exist. Numerous kinds of mammals, reptiles, and birds also
perished.

Many explanations have been offered for the dinosaurs' complete and total demise. Devastating plagues, magnetic-field rever-
sals, increased tectonic activity, as well as severe climatic variations
triggered by stray asteroid impacts or supernova explosions have
all been proposed, each with some merit, though none entirely
convincing. Out of seeming desperation, some researchers even
joke that the dinosaurs died of constipation, since a large variety
of oily plants on which they probably feasted also became extinct
at about this time. For whatever reason the dinosaurs perished,
dramatic change of some sort was surely responsible. It would be
useful to continue the search for the cause of their extinction, for
there's no telling if that sudden change might strike again. As the

dominant species on Earth, we are the ones who now stand to lose the most.

Despite our knowledge of dinosaurs, no human ever saw one. Furthermore, their remains lay hidden for nearly a hundred million years before *Homo sapiens* discovered their prior existence only about a hundred years ago. It's hard to believe, but our great-great-grandparents could have known nothing of the dinosaurs. In fact, just recently did we come to realize that we probably wouldn't be here had these grand hulks not died out. Only when the dinosaurs disappeared did the spectacular rise of the mammals — including human beings — begin.

The onset of the Cenozoic age, nearly a hundred million years ago, saw the appearance of an almost entirely new cast of characters. Dinosaurs were completely gone. The earlier reptilian dominance over the mammals had been totally reversed. And the planet had returned to its predinosaur tranquillity, even becoming somewhat desolate in places. Clearly, mammals had taken over the world, though they apparently did so by default. We might say that the meek had indeed inherited the Earth.

Fossils as recent as fifty million years old show that most mammals had small brains, large jaws, clumsy and inefficient feet and teeth. Still, they were freely multiplying, swelling in numbers and diversity. As always, change was rampant. Ice ages came and went; continents split and drifted. Life forms were constantly fine-tuning their day-to-day routine for better survivability. Accordingly, many of these early mammals passed into extinction, to be replaced by better-adapted stocks.

In a relatively short time, the mammals had evolved into an amazing assortment of creatures. Some forty million years ago, the ancestors of such modern mammals as the horse, the camel, the elephant, the whale, and the rhinoceros, among others, gradually ventured forth, though in shapes almost totally unrecognizable compared to their descendants of today. Most grew in size and improved their overall performance between forty and twenty million years ago.

∞

Many species have come and gone on planet Earth. Some were inconsequential organisms, while others ruled the land, sea, and air. For hundreds of millions of years, there has been a steady parade of new creatures, many of which led in turn. Yet, only the latest of these dominant life forms — men and women — are able to learn about those that went before. Humans alone have been able to unearth and understand the absolutely amazing chronicle of now-extinct and terribly bizarre life forms once prevalent on our planet.

What sense can be made of the vast array of past and present life on Earth? Is there any unifying logic linking it all? Well, classification is the first step in any attempt to discover the underlying causes of the abundance and diversity of life on our planet.

All current life as well as fossilized remains of ancient life can be broadly classified as bacteria, plant, or animal. These classes can, in turn, be further divided into different species, a subclassification generally used to imply not only structural similarity but also an ability to mate and produce fertile offspring.

Actually, understanding involves more than just categorizing life forms. Real creatures do not always match what is expected of a species. In other words, individual species usually show small, though noticeable, variations from their "ideal" classification — slight alterations from the standard specimen to which each individual organism may be compared. This is true of all species, whether they're now living or fossilized.

As with many aspects of matter in the Universe, subtle changes in life forms contain the key to our understanding of how life has developed throughout the course of time.

The theory of biological evolution, independently conceived by the nineteenth-century British naturalists Charles Darwin and Alfred Wallace, can account for two outstanding features of the fossil record: First, living systems have generally become more complex with time; and second, members of all species show some variation from their "ideal" category.

These two facts clash head-on with the age-old assumption that nature is immutable. Like Copernicus four centuries before, and Heraclitus twenty centuries before that, Darwin and his associates faced the same kind of opposition championed by lingering Aristotelians who refused to concede that Earthly things change. But a static theory of life is simply untenable. Everything changes with time, life included. The only feasible explanation is a dynamic, evolutionary one.

The central tenet of biological evolution maintains that living things change, some for the better, others for the worse. Those that survive for lengthy periods are often drastically modified, sometimes becoming entirely new species. Some species become extinct; others arise anew. Organisms of similar structure have similar ancestry and are closely related. Those with very different structures have accumulated these differences over long periods of time and are therefore now only distantly related.

Biological evolution is not faith. It's fact. The fossil record no longer leaves room for any reasonable doubt that evolution does happen. The "what" aspect of evolution is backed by data. The "how" aspect, however, is less clear.

The precise mechanisms underlying biological evolution compose a theory — a theory developed to account for a group of facts. It's the result of the scientific method: Observations were made of the fossilized remains of life; a hypothesis was constructed to explain those facts; and subsequent experimentation has served to strengthen and revise the intricacies of the theory during the past century.

What is the mechanism of biological evolution? How does it work? The prime mover is the environment, the macroscopic conditions surrounding all living things. Temperature, density, foodstuffs, air composition and quality, in addition to natural barriers such as rivers, lakes, oceans, and mountains, are all influential environmental factors. Other factors are more subtle, such as personality clashes, neighborhood conditions, and scores of exceedingly complex sociological predicaments. Complicating the situation further is the fact that these environmental pressures constantly change, albeit slowly. Biological evolution dictates that all life forms respond to their changing environment, inhibiting some traits while promoting others, but in any case yielding an immense diversity of species throughout the course of time. Changes — in the environment, and in life — occur as a rule, not as an exception.

Observations show that, although all species reproduce, few of them display gigantic increases in population; the total number of any one species remains fairly constant, there being no dramatic explosions of offspring. Furthermore, the process of reproduction is never perfect; offspring in each generation are never exact copies of their parents. The implication is that most offspring never survive to reproduce. All life must struggle and compete in order to endure.

Natural selection, an expression coined by Darwin himself, is the mechanism that guides life's evolution along the arrow of time. Recognizing that most members of a species exhibit some variation from their ideal standard, Darwin suggested that organisms having a variation particularly suited to their environment would be most likely to survive. They are quite naturally selected to live. By contrast, those organisms having unfavorable variations would be most likely to perish. They are naturally selected to die. Only those life forms able to adapt to the changing environment would survive to reproduce, thereby passing these favorable variations or traits to their descendants.

In successive generations, advantageous traits become more pronounced in each individual. Not only that, but the numbers of individuals possessing favorable traits also increase. Favored individuals generally produce larger families, as they and their offspring have greater opportunities for survival. Their favored descendants multiply more rapidly than those of their less advantaged neighbors, and over many generations, their progeny replace the heirs of individuals lacking the desirable trait.

Natural selection truly does smack of the well-known phrase "survival of the fittest." It literally molds life forms. With the passage of sufficient time, the action of natural selection can considerably alter the shape, disposition, and even the existence of individuals; old species disappear in response to changing conditions, while whole new ones arise.

Natural selection cannot be easily observed at work; long passages of time are required to note a distinct variation in any population of a species. Some success has been achieved using laboratory conditions that mimic those of nature. Like the origin-of-life experiments, these are simulations that attempt to study the adaptation of life to a changing environment. The results support the theory of biological evolution via natural selection. What follows is an example of one such experiment.

Two sets of field mice, one set with dark fur and the other with light, were let loose in a small barn along with an owl. The straw and ground cover were chosen to match closely the dark coloring of one set of mice. This then gave the darker-colored mice an environmental advantage; the lighter-colored mice were clearly at a disadvantage. At the end of carefully controlled experiments, the owl had indeed captured many more lighter-colored mice. When the ground cover was lightened — corresponding to an environmental change granting the lighter-colored mice a greater opportunity to survive — the results were reversed; the owl readily captured the darker-colored mice. This is an example of how small variations in one species can grant a competitive advantage. Whichever mice survived did so because they were better adapted to the environment. They were naturally selected to live and thus to reproduce their kind.

Some examples of natural selection have also been noticed in nature's outdoor setting where an environmental factor changed exceptionally fast. For instance, more than a hundred years ago, the bark of all birch trees was nearly white, enabling light-colored moths to blend nicely with their environment. In their struggle to survive, they prospered around the trees against whose bark they were nearly invisible. Their darker-colored relatives lacked this competitive advantage, because they stood out clearly against the white bark, allowing birds to snare an easy meal. By the turn of the last century, however, the bark of birch trees near some manufacturing cities had become heavily soiled with the sooty pollutants of the Industrial Revolution. This environmental change — rapid by nature's standards — had removed the benefit previously enjoyed by the lighter-colored moths. The result is that few pale moths remain today, at least near industrialized areas; instead, grayish moths now possess the advantage of camouflage, enabling them to prosper, mate in peace, and freely reproduce. This is an example of how simple variations — in this case color — serve to guide natural selection within a changing environment. For some moths, in fact, a small change became a matter of life or death.

The common housefly presents a second example of how some members of a single species can adapt to a changing environment, granting them a better chance to be naturally selected for survival. Originally, the pesticide DDT was successful in killing houseflies. Few flies were able to adapt to the sudden environmental change caused by the chemical DDT in the air. A small minority, however, managed to survive because they possessed a chance variation or trait that made them immune to this chemical. Environmental change didn't faze these oddballs; it enabled them to reproduce freely and thus pass the advantageous trait on to their descendants. Within a decade, the offspring of the oddball survivors outnumbered the original type of fly. Accordingly, DDT has grown less effective over the years. Now most houseflies have inherited a resistance to DDT, and the pesticide is useless against these flies. The chemical DDT didn't give this resistance to the flies; rather, it provided an environmental change enabling natural selection to go to work. To survive as a species, the housefly had to be able to adapt to the changing environment. Those that managed, survived. Those that were unable to, are long gone.

These last two cases are examples of evolutionary responses to environmental changes induced by humans — a whole new aspect of evolution whereby technologically equipped beings play the role of nature.

Over long periods of time, chance variations in living things can accumulate. Hair color, eye color, size, shape, appearance, and a host of other traits all change as nature naturally selects for survival those life forms best adapted to the environment at any given time. Eventually, some life forms come to differ considerably from members of their original species. In this way, the environment helps new species to evolve from old ones.

For example, members of a single species may be disrupted by some physical modification in the environment, say a new river that's gradually rerouted through an area inhabited by a species of butterflies. Should the river become wide enough to act as a physical barrier that cannot be crossed, butterflies on one side would then be prohibited from mating with those on the other side. In this way, the two populations of butterflies become completely isolated, so that over long periods of time, they can become considerably different. Should the barrier be removed — if the river dries up, for instance — then the two populations would be able to intermingle once again. Provided they were separated for a long enough time, however, they will be unable to interbreed, meaning that two new species of butterflies exist where previously there was only one. Each new species will furthermore stake out its own claim or fill a separate niche, thereby coexisting peacefully within the new environment.

Environmental disruptions of this sort often guide the transformation of a single species into two or more species. Known as speciation, it's the mechanism behind the diversification of all life.

Consider an actual case study involving recently upthrust mountains and eroded ravines. Two distinctly different populations of squirrels live on the north and south rims of the Grand Canyon. The Kabib squirrels of the north rim have black bellies and white tails, whereas the Abert squirrels of the south rim have white un-

derparts and gray tails. Both feed on pine-tree bark growing only
on the kilometer-high plateaus. The two populations are now
separated, and supposedly have been for thousands of years, by the
intensely hot and dry environment in the canyon. But they have
so many similarities that it seems safe to assume that their ancestors
were once members of the same species.

There are many other examples of two or more slightly different
species, coexistent but clearly isolated, and presumably sharing a
common ancestral heritage. Biologists find more of them every day,
including members of species that aren't even separated by a physical barrier, but that for one reason or another do not interbreed.

What is it that alters living systems to make members of a single
species occasionally unable to interbreed? Basically, the microscopic gene is the culprit, for it's the genetic code that dictates if
and how life forms reproduce. Pioneered about a hundred years
ago by the Austrian monk Gregor Mendel, the subject of genetics
has become a lot more complex than he could ever have imagined.
Darwin himself would probably be surprised at the microscopic

roots of biological evolution as we know it today — a modern synthesis of Darwinian and Mendelian ideas, often referred to as neo-Darwinism.

Still, what causes the genetic alterations? What factors contribute to the similarities and differences in organisms? In short, what is the origin of all the myriad variations seen throughout the living world?

Hereditary error is a major factor promoting the evolution of living systems; it is in fact a prerequisite for evolutionary change. Note that we say hereditary *error*, not heredity itself, which is an agent of continuity, not change. Heredity is that biological phenomenon ensuring the preservation of certain traits in future generations of a species. Otherwise, the basic life processes and body organs of each and every organism coming into the world would have to be created from first principles. Normally, chemically coded instructions of the DNA molecules enable cells to duplicate themselves flawlessly millions of times. But, occasionally, mistakes do occur at the microscopic level. Not even genes are immutable. Everything changes.

For reasons not completely understood, a DNA molecule can sometimes drop one of its bases during replication. Or it may pick up an extra one. Further, a single base can suddenly transform into another type of base. Even such slight errors in the DNA molecule's copying process mean that the genetic message carried in the DNA molecule for that particular cell is changed. The change doesn't have to be large; even an alteration in one nucleotide base out of millions strung along the DNA molecule can produce a distinct difference in the genetic code. This, in turn, causes a slightly modified protein to be synthesized in the cell. Furthermore, the error is perpetuated, spreading to all subsequent generations of cells containing that DNA.

Microscopic changes in the genetic message, called mutations, can affect progeny in various ways. Sometimes the effect is small, and newly born organisms seem hardly any different. At other times, mutations can affect a more important part of a DNA molecule, inducing considerable change in the makeup of an organism.

And, at still other times, a single mutation can rupture a DNA molecule severely enough to cause the death of individual cells, or even whole organisms.

Mutations are responsible for differences in hair color, eye color, body height, finger length, skin texture, internal structure, individual talents, and numerous other characteristics among a population of life forms of any given species. Virtually any aspect of the life of any organism can be modified by genetic mutations. Such mutations provide a never-ending variety of new kinds of DNA molecules.

All mutations are not detrimental. Most of them do indeed create traits inferior to the previous generation's — especially in many of today's highly evolved and exquisitely adapted organisms. But some mutations are favorable, and serve to ameliorate the life of an individual. These can then be passed on to succeeding generations, making life more bearable for members of that species. Beneficial mutations act as the motor of evolution, steering life forms toward increasing opportunities to adapt further to the ever-changing environment.

What causes genes to mutate? Why do some DNA molecules occasionally replicate differently, though they may have copied themselves exactly for millions or even billions of previous cell divisions? Frankly, the precise causes are unknown, for they apparently arise from chance, indiscriminate events. Biologists have undertaken laboratory experiments with cells, and have succeeded in increasing the numbers of mutations by artificial means, thereby helping to unravel the ultimate reasons for genetic change. The results so far show that the easiest way to enhance gene mutations is to treat reproductive cells with external agents.

Three of the most important mutation-inducing agents revealed in the last few decades are temperature, chemicals, and radiation. When cells are heated, or treated with industrially generated nerve gas or chemical drugs, mutations are clearly enhanced. In addition, ultraviolet and X-ray radiation seem to be particularly striking causes of genetic mutations. Such radiation has been present on Earth, in one form or another, since the origin of our planet.

Radioactive elements embedded in rocks, cosmic rays bombarding Earth from outer space, and solar ultraviolet radiation reaching the ground all serve to prove that life evolved in a radiation-filled environment.

Generally, there's nothing wrong with immersal in radiation. We and other life forms probably wouldn't be here if the motor of evolution had not speeded change; without radiation, life itself might not have progressed beyond the primitive, unconscious, unicellular organisms drifting in the oceanic slime. Of some valid concern now, however, is the fact that human inventions such as atomic bombs, nuclear reactors, and medical devices also release radiation. Intense doses of radiation can kill directly, though more subtle doses cause changes in the reproductive cycle that are then passed on to future generations. It's not clear that these human-induced mutations are in all cases harmful, but, in the absence of evidence to the contrary, a healthy degree of skepticism is surely warranted.

We must not risk damaging the refined work of several billion years of organic evolution, for the long series of accidental changes that evolution represents can never be repeated.

∞

A hundred years ago, the concept of biological evolution was intellectually and morally shocking. Few people accepted it; even many scientists of the late nineteenth century failed to embrace it. The problem was not really the idea of evolution. Surely evolution occurs, and people of a century ago knew it. Fossils were already abundant then, and agriculturists had for centuries bred crops and livestock in the successful effort to develop healthy, disease-resistant strains. The real problem was that people were disturbed to hear that humans had anything in common with a bunch of apes. When it comes to hypotheses involving humans, vanity often surfaces like an irrepressible force. Because of this same vanity, apparently, minor segments of our twentieth-century civilization still refuse to accept the basic tenets of biological evolution.

Scientists now possess a combination of fossil discoveries and behavioral studies virtually proving that, of all species of life now on Earth, the chimpanzee and gorilla are our closest relatives. Humans haven't descended from apes, a common misunderstanding. Rather, modern science stipulates that apes and humans have some common characteristics, and thus a common ancestor. We shouldn't be able to identify that ancestor from among the presently living creatures on Earth, for genes and environment change over the course of millions of years. But such an ancestor should be part of the fossil record.

To discern our most recent ancestors, and thereby trace the ways and means of relatively recent biological evolution, scientists rely heavily on the fossil record. Teeth and skull bones have accounted for the majority of fossil discoveries ever since people began digging around for artifacts in Earth's rubble two or three centuries ago. Teeth are the most enduring part of any life form because of their extraordinarily hard enamel. Skulls are the most recognizable part, largely because they are more noticeable than arm or leg bones among sticks and other common ground litter. Careful study of these and other bone fragments has now enabled researchers to arrive at a consensus for the lines of ascent culminating in humanlike creatures. The idea goes like this:

Early in the Cenozoic age, not quite a hundred million years ago, squirrellike mammals were looking for ways to increase their chances of survival within a rather harsh environment. These were insect-eating creatures living primarily on the ground. The dinosaurs were gone by that time, but life on the ground still presented problems for these ancient mammals. The fossil record implies the existence of relatively large creatures who no doubt survived at their expense. Fortunately, though, sporadic mutations and a constantly changing environment granted some of these small mammals a chance to change their living patterns.

Many mammalian species at this point invaded the trees. They were undoubtedly searching for more food, while trying to escape the fierce competition prevalent on the ground. Some species found the trees an even rougher environment and thus became extinct. Other species discovered the trees to be to their liking, surviving famously. A few, like the tree shrews of Southeast Asia, still thrive today. Their traits just happened to be adaptable to life in the trees. In fact, the trees constituted a whole new niche, helping to transform these creatures from ground-dwelling and insect-eating mammals to tree-dwelling and banana-eating prosimians. These protomonkeys are the least advanced members of the order of primates, a zoological category to which apes and humans belong.

Fossils document seemingly endless refinements as each of the most successful early prosimians ferreted out the very best available niche. With time, over generation after generation of natural selection, paws gradually transformed into hands. Stubby claws eventually became flexible fingers. And the opposable thumb took shape as the environment promoted its prominence as a superior tool for maneuvering among the branches. Remember, these were not anatomic changes that occurred as individuals grew during single lifetimes. Instead, these were genetic changes endured over the course of millions of years. Favorable mutations eventually gave those prosimians having good balance, keen eyesight, and dexterous hands and fingers a naturally increased opportunity for survival within their newly discovered tree-based environment. Those better adapted to this environment could reproduce more efficiently, thus passing these favorable traits on to their generations of offspring. Some creatures more than others excelled in jumping, leaping, swinging, clinging, and food-gathering. The result, again documented by the fossils, was widespread speciation resulting in the appearance of legions of novel tree-dwelling species.

The evolution of accurate sight was a particularly important development. Trees are, after all, three-dimensional, unlike the flat two-dimensional ground. The advantageous trait of smelling on the ground gave way to that of seeing in the trees. Fossils show that, over the course of millions of years and generation upon generation, mutations gradually brought the eyes of some of these

tree-dwellers around toward the front of the head, thereby producing binocular, stereoscopic vision. With the eyes at the sides
of the head, two independent fields of view result, much as we
might perceive upon placing our nose and forehead against the
edge of an open door. The gradual shortening of the snout and the
slow displacement of the eyes toward the front granted some early
prosimians an overlapping field of view and thus more sophisticated vision. Depth perception, in particular, enabled them to
gauge distances among the branches. Clearly, these distant ancestors of about fifty million years ago had distinct advantages in the
struggle for survival. They had become monkeylike creatures. A
major new evolutionary path had originated.

Fossil findings also disclose that some species of monkeys gradually became larger. Again, a single generation of a given species
didn't suddenly balloon in size. Rather, sporadic mutations in their
DNA molecules, spread over scores of generations, granted larger

monkeys some advantages in the competition for survival. For example, larger, more aggressive males have clear superiority over smaller ones in the sexual competition for females. Also, bulky bodies usually provide additional protection from predators.

On the other hand, large size is not a wholesale advantage. Sheer mass brings some problems, too. Bigger monkeys, for instance, find it troublesome to hide, and they also need more food to survive. There are both advantages and disadvantages associated with all genetic changes; only when the advantages outweigh the disadvantages is there an enhanced opportunity for living.

The ability to grasp a branch securely while simultaneously extending an arm to secure food also provided an obvious advantage at the time. Those species unable to cling well enough to hold on plunged, died, and became extinct. Those species without long enough arms to reach the food likewise starved, died, and became extinct. The clear advantage was had by those prosimians able to coordinate clinging and grasping simultaneously. Of course, being smart enough to repel attacks from an array of enemies didn't hurt. The tree of forty million years ago thus became a fairly comfortable niche for some species of monkeys. These well-adapted creatures could probably have remained in the trees indefinitely if a problem hadn't arisen. Fortunately for us, change stirred; otherwise we wouldn't be here.

Leisurely life in the trees evoked fundamental trouble. The ancestral monkeys of forty million years ago were so comfortably accustomed to their tree-dwelling environment that they multiplied faster than many other species stuck in harsher environments. Time not spent trying to survive can be most agreeably used trying to reproduce. The result was probably a population explosion, the type of crisis that's inevitably followed by a food shortage. Consequently, some of the prosimians survived only by utilizing their limited ingenuity to discover new sources of nourishment. For some, that meant leaving the trees and returning once again to the ground.

Some species of monkeys remained in the trees. Most of these eventually became extinct, though some survive in altered form today. Baboons, gibbons, orangutans, and numerous other modern tree-dwelling creatures are the descendants of the well-adapted monkeys that elected to remain in the trees. Those successful prosimians that came down out of the trees inaugurated a whole new evolutionary track. That was the path toward the ape and human species.

At first notice, it seems foolish for the prosimians to have taken up residence in the trees about seventy million years ago if some of their descendants were just going to have to depart those trees several tens of millions of years later. But a critically important change occurred while they were aloft: They evolved and adapted to a very special environment. When the tree-dwelling prosimians returned to the ground, they were equipped with qualities that probably would not have been naturally selected had they stayed on the ground. With their manual dexterity and binocular vision, among other assets, they were far more advanced than any other type of life then on the ground.

Had our prosimian ancestors not taken refuge in the trees, they might never have experienced the need to foster these exquisite capabilities, many of which we now use, for example, while writing and reading books like this one. The forty-million-year detour into the trees was well worth the time and effort. By thirty million years ago, the ground-dwelling creatures had become dominant. They hung around forests, not venturing too far from the protection afforded by the trees. They foraged for food primarily on the ground, gradually evolving larger brains and some degree of bipedalism, as well as the rudiments of the technical and linguistic skills needed for the development of culture. Our ancestors of the past thirty million years encountered few roadblocks in the final rush along the evolutionary paths that led to the variety of peculiar animals now inhabiting our planet, including street-wise humans.

CULTURE

Intelligence to Technology

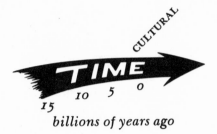

billions of years ago

ONE OF THE MOST REMARKABLE ASPECTS of life is an awareness of its surroundings. Unlike nonliving matter, life can monitor impressions from, and respond to, the outside world. Through life's various senses — hearing, seeing, smelling, touching, tasting — organisms acquire and file enormous amounts of information. The extent to which beings are successful in doing so depends largely upon their "complexity." Organisms manifest this complexity in terms of one exquisite piece of matter — the brain. The brain is the central clearinghouse of all behavior.

As you read these words, matter within your skull is full of electrical activity. Silently, though efficiently, millions of nerve cells pass messages back and forth within your brain. They guide your eyes along this printed line. They quickly scan the shapes of the letters. And, by matching them against memory, they allow you to recognize words.

Nerve cells constantly interchange signals within our brains, ordering our hearts to beat, our lungs to pump, and your hands to stand ready to turn this page. The body's nervous system, of which the brain is the paramount part, controls all mental and physical activity. In fact, every thought, feeling, or action begins in the brain. All human behavior is controlled by it.

Most interesting of all, these silent, unfelt activities inside your head make you aware that you're now reading about them. Amazingly, the brain can contemplate the brain.

The human brain is the most complex clump of matter known. It's nature's most tantalizing, talented, and versatile creation — the ultimate example of the extent to which matter has evolved in the *known* Universe.

∞

The one-celled amoeba is the most primitive form of life known in today's world. Roughly halfway between an atom and a human, it has poor awareness and coordination. It generally responds only at the point stimulated, communicating the information sluggishly through the rest of its body. Though amoebas have developed a crude nervous system, more agile living things surely need quicker internal communication.

Other single-celled creatures have succeeded in developing primitive intercom systems. The microscopic paramecium, for instance, possesses an array of oarlike hairs that enable it to move rapidly through water. The "oars" must act in a coordinated way, for if they functioned independently, the paramecium would never make any progress. They are regulated by microscopic nerves known to respond to a chemical emitted within the cell. In this way, messages can be transmitted swiftly and precisely from one part of the cell to another.

The paramecium clearly has an "intelligence" superior to that of the amoeba. It has better coordination, and it has a memory of sorts. An amoeba searches for food by drifting into water-plant algae; if it finds none, it often repeatedly gropes toward the same alga, even though that alga has no satisfactory food for the amoeba. The amoeba has no memory. The paramecium, on the other hand, having found no food near one alga, will back off and seek resources in a different direction — it momentarily retains traces of experience.

Compared to the amoeba, then, the paramecium is a genius. But

it's a genius operating in a watery world no more than a few milli-
meters in extent. It's unaware of anything beyond. No single-celled
creature can be much smarter, for it can develop no further.

Despite the biochemical complexities of a single cell, it can
boast only the simplest intelligence. To become smarter — to develop an intricate nervous system — a single cell would require
elaborate sense organs to inform it, as well as developed muscles
to implement its instructions. Why can't there exist, then, larger
cells incorporating these added luxuries — perhaps equipped with
miniature hands, eyes, and brain? The answer is that single cells
cannot become much larger than the 0.0001-centimeter creatures
described above. Should they try to do so, their surface areas would
increase with the square of their size (1, 4, 9, 16, . . .), whereas
their masses, which must be fed through the cells' membranes,
would increase as the cube of their size (1, 8, 27, 64, . . .). So cells
cannot become too large, lest they starve. The basic smarts of
single-celled life forms are thus limited; their physical size prevents
them from developing the many and complex organs necessary for
higher intelligence. Mutations have undoubtedly helped them try
every conceivable means to do so for the past three billion years,
but they have failed.

The road to greater intelligence requires many cells. A haphazard accumulation of many independent cells won't do the trick,
however; clusters of a million independent cells are no more intelligent than one cell. Consider the sponge, for instance, not unlike
those harvested for use in our bathtubs. Though the sponge is
clearly multicellular, most of the millions of cells within it act
independently. A sponge has no central nervous system, and thus
is really not much more intelligent than an amoeba. For some
reason, it failed to profit by its multicellularity. As a result, it has
produced no higher forms of life. The sponge is an example of
a life form that long ago reached an evolutionary dead end.

What was needed was a favorable mutation allowing an accumulation of many cells to work together as a community. Inter-
active, multicellular organisms have some clear advantages, not the
least of which is that they avoid the surface-volume difficulty just

mentioned. More important, groups of cells within a multicellular organism can develop specialized functions. This division of labor was one of nature's greatest inventions. One group of cells can become highly sensitive to foods; others, more efficient in carrying oxygen; still others, tough muscular entities or protective skin casings. The net result is that each group of cells within a multicellular organism becomes more skilled in one capacity and less so in the rest. Accordingly, the total intelligence of an organism is enormously increased as cells, working as a team, become better able, not only to protect themselves from predators, but also to obtain the food necessary for survival. These were the first steps toward symbiotic society.

The animal called hydra is a good example of a multicellular organism that did evolve considerable intelligence. No larger than a toothpick, the modern hydra resembles a stalk of celery, being closed at the lower end and raveled into writhing appendages at the upper end. In contrast to any sponge, the hydra can move its entire body in coordinated fashion to, among other things, avoid danger and seek food. In short, the cells within the hydra can communicate. And communication is the essence of organized intelligence.

Cells able to communicate — nerve cells — probably formed originally near the surface of multicellular animals such as the hydra, or, more realistically, hydralike progenitors. Being exposed, these cells had the greatest opportunities to sample their environment. But being near the surface also made them more vulnerable. Thus, mutations and natural selection favored those hydralike ancestors having deeply rooted nerve cells. These cells gradually retreated to the interior of the organisms but kept their link to the environment by sprouting expendable tentacles that reached the surface and beyond. These miniature octopuslike tentacles are the dendrites of modern neurons, the special cells that communicate information in intelligent beings.

As evolution proceeded, the bulk of the neurons themselves retreated ever deeper within multicellular organisms. Eventually, the buried neurons formed a coagulated entity of nerve cells, the

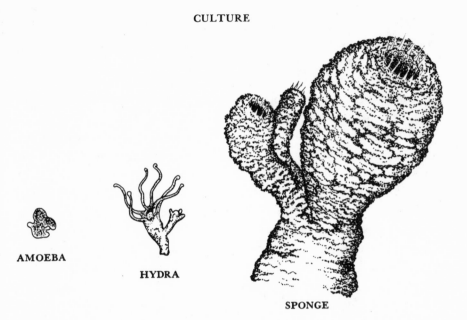

AMOEBA

HYDRA

SPONGE

first and most important step in the development of a central nervous system. Neuron clustering was one of the greatest of all evolutionary breakthroughs. Once that barrier was crossed, sometime around a billion years ago, our hydralike ancestors, as well as other sophisticated organisms like them, were on their way to generating all of Earth's brainy animal life forms, including humans.

What does the fossil record say about the evolution of the brain? It shows a clear evolutionary development of the central nervous system, with organisms branching out in every available direction, moving steadily toward greater complexity. Most of the branches, however, represent organisms that either became extinct long ago or survived only as dead ends. Apparently, at some point in their development, they met some insuperable biological obstacle to progress. Some that persist without progress include the amoeba, paramecium, sponge, and hydra, as well as worms of all sorts. These are the invertebrates, skeletonless organisms, many of which are skilled and crafty in their own domain. Spiders, for instance, are marvelously accomplished performers within their own environment; their nervous systems are complex and effective, their sense organs even more varied and subtle. Bees, wasps, ants, and moths also have superb bodies for dealing with their world. Some — especially the bees and ants — even have sophisticated social organizations incorporating simple symbolic communication.

All these invertebrate animals, however, have reached evolutionary dead ends. They're trapped in an endless cycle of perfected day-to-day routines. Fossilized spiders of a hundred million years ago show little variation from their modern descendants. Bees in the bushes, spiders in the shed are, in a sense, living fossils.

Invertebrates are successes and failures at the same time. They are fabulously successful within their own small environment. The deerfly, for example, is judged the fastest animal; the flea can jump a hundred times its own height. Successes certainly, for the invertebrates dominated Earth for nearly a half-billion years. But failures too, because they neglected to develop the vertebral column of bones so conspicuous in fish as well as humans — bones that form the spinal column and skull.

Vertebrates are a minor offshoot from the vast, teeming world of invertebrates. Humans and our fellow vertebrates (skeletonized fish, reptiles, and mammal relatives) are an exception to the great invertebrate failure.

As humans, we take brains for granted. But the vast majority of animals are invertebrates, and thus can have no true brain, no centralized nervous system. As such, they cannot be creative, adventurous, or visionary. Brains are the exception, not the rule.

The vertebrates' foremost characteristic is their central nervous system. Even so, many of them were apparently unable to utilize sensory and motor organs to full capacity. A vast array of fish, amphibians, and reptiles, including modern versions of many birds, lizards, snakes, crocodiles, turtles, and many other vertebrates, dead-ended long ago. A good number became extinct, and even the survivors seem to have been unable to decide on a division of authority between the "sight" and "smell" neurons.

Skulls of primitive fish have been reconstructed in some detail

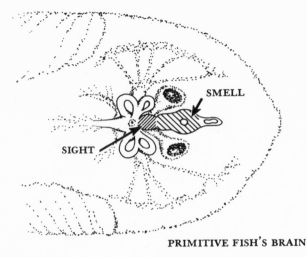

PRIMITIVE FISH'S BRAIN

from the fossil record. These fish lived several hundred million years ago, and were among the simplest known true vertebrates. Though clearly crude, their brains nonetheless contained all the essentials present in modern fish as well as humans. The clustering of smell organs caused a bulge toward the snout, and was the precursor of the much larger cerebral hemispheres in humans. Similarly, their eyes created another bulge farther back, precursor of our occipital lobe, on which we "see" images projected at the rear of our brain. Lateral-line organs also branched out to the side, precursors of our cerebellum, where our body movements are refined and coordinated. These ancient sense organs, though not in themselves rivaling those of some modern invertebrates, were employed more efficiently because of their connection to a unified central nervous system.

The development of specialized sense organs, coupled with the integration of those organs with the centralized brain, contributed to the intellectual dominance of the vertebrates over the invertebrates.

Sight certainly played a major role in the advancement of these early vertebrates, as the fossil record documents the development over time of a relatively large sight brain. Mutations simply gave an advantage to certain species of fish, enabling them to utilize enhanced eyesight to move, survive, and procreate better in the water. The sense of sight didn't rule unchallenged, however. The sense of smell remained a keen rival in the ever-refining development of Earth life several hundred million years ago.

Interestingly enough, when amphibians transferred from the sea to the land, the flood of sight data probably overwhelmed even the sophisticated brain of these rather advanced vertebrates. Smell input, on the other hand, was still within the range of such a brain. For the first amphibians, smell was of more practical use than sight. The fossil record suggests, in fact, that the occipital lobe shrank, while the cerebral hemispheres expanded with the march of time. Gradually, the sense of sight regained greater usefulness as the brain of the mammals grew larger through continued mutations and natural selection; the multitude of out-of-the-water images no longer saturated once oceanic eyes. The larger brains of the mammals were then able to cope with the full world of sight as well as sound. Those creatures having more sophisticated brains were clearly better suited to survive in a changing terrestrial environment.

The much larger human cerebral hemispheres are indeed derived from the ancient smell brain, but the preeminence of this sense was long ago surpassed by sight, sound, and other general sensations.

The final step in the evolution of the brain occurred in the mammals. There were once again many evolutionary dead ends — mutations that simply didn't give much advantage to a particular species. However, other mutations did alter for the better the traits of some organisms, thus granting them a distinct advantage in the overall struggle for survival.

We've noted how favorable mutations, over millions of years and generations upon generations, caused the development in some mammals of longer arms and gripping paws for leaping, swinging, and reaching food. Other mutations also gradually relocated the eyes of the early prosimians from the side to the front

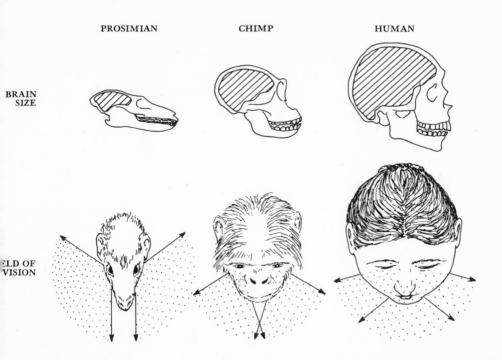

PROSIMIAN   CHIMP   HUMAN

BRAIN
SIZE

ELD OF
VISION

of the head, thus granting binocular, stereoscopic vision. The shortening of the snout and the movement of the eyes toward the front granted the early primates an overlapping field of view and thus three-dimensional sight. The eventual development of longer arms and more dexterous hands combined with the more accurate eyesight to give these protomonkey ancestors distinct advantages in the struggle for survival.

All the while, the cerebral cortex expanded, enabling mammals to develop increasing sense sophistication. The combined eye-hand-brain system was a powerful tool for not only enhancing survival but also evolving intelligence. The culmination is the seemingly limitless manual skills of *Homo sapiens*.

What about ourselves? It's natural to wonder about the specific evolutionary path that led from our ancient ancestors to contemporary humans. Where did we come from? How did we get this way? What were the circumstances that led to our decidedly odd body shape? Specifically, what factors led to the development of our fabulous attributes of thinking with our brain, talking with our mouth, walking on our hind legs, constructing with our hands, and seeing with binocular vision?

Who are we? Not what is our Sun, our planet, or life itself. But who or what are these twentieth-century human beings inhabiting Earth? We all ask ourselves that question at one time or another. It's one of the most profound and interesting issues.

In a nutshell, each of us seems to be the product of many ancestral life forms — a cluster of genes inherited from all of them, and shaped by an environment that's partly ours, partly our parents', partly our parents' parents', and so on.

Tracing back a thousand years, each of us would have had more than a million ancestors, all alive at the same time. They were probably spread over much of the world, living in a diversity of environments. Going back another few thousand years, some of our ancestors could well have been members of the ruling class of ancient Egypt or Babylonia. But the bulk of our ancestors were probably slaves or peasants. Chances are good that they could neither read nor write. They were probably ignorant, superstitious, and cruel — primitive agriculturists at best. Few of them would have touched metal or operated a wheel. By modern standards, most of our ancestors of several thousand years ago were savages. They survived largely by hunting and gathering. It's hard to relate to them, but modern science suggests that we must. Evolution stipulates that we carry some of their genes in our bodies. Portions of our features, shapes, desires, and attitudes, as well as our outlook and philosophy, trickled through from the genes of our ancestors, molded in part by the environment in which they lived.

Answers to the fundamental questions, then, are evolutionary ones. They're answers that permit us to relate ourselves to all humankind, indeed to all living things. If we can find these answers, then perhaps we can determine who we really are, how it is that we can walk upright, fashion tools, and converse meaningfully, as well as think, speculate, and exhibit curiosity about ourselves and our Universe.

Long ago, our distant ancestors possessed none of these attributes. They were not human. They were small-brained beings populating forests. Somehow they gave rise to humans. Somewhere in our ancestral past, there are links joining creatures that clearly were human with creatures that clearly were not.

Fossils allow scientists to sketch the paths of evolution, and to understand the fine lines of distinction between various closely related species. Fossils of recent times are usually well preserved, enabling researchers to document evolution with a reasonable degree of assurance. Not surprisingly, the older fossils are in poorer condition, often in pieces, and sometimes hardly recognizable. Reassembling the pieces is much like working a jigsaw puzzle. Trying to understand where and when the reconstructed fossils fit into the evolutionary line of ascent is another kind of puzzle.

The entire effort of unraveling human genealogy is often likened to the restoration of a gigantic mural painted over the course of millions of years. To the right, where the scene has been sketched in recent times by modern humans, the message is reasonably clear. Toward the left, the mural is soiled, peeling, and generally deteriorating. Painted long ago by our ancestors, the mural cannot be easily cleaned of dirt, nor can it be easily repaired to reveal the message once that dirt has been removed. Like any restoration, the process is a slow and painstaking one, done very deliberately to avoid destroying the mural and thus the message.

Anthropologists and archeologists examining old fossils need the virtue of patience in addition to an inquiring though unopinionated mind. It's very much like detective work, trying to ferret out an entire story based on just a few hints. The solution here, however, means more than closing another crime case. It means knowing just that much more about the origins of human beings, a superbly satisfying objective.

The late nineteenth century was an exciting time for geologists — a time full of discoveries and revelations about our planet. Field excursions and archeological excavations were only then becoming popular. Scientists began recognizing that the ground below their feet held clues to the nature of Earth, and especially to former life on Earth. Among the most fascinating results were the first discoveries that there once existed such life forms as dinosaurs, the former rulers of our planet. Equally intriguing were ancient stone axheads and crude implements of all sorts uncovered along rivers and inside caves of western Europe. These were primitive tools, but tools nonetheless. Geologists found that many of them were nearly a hundred thousand years old. The question arose: Who were the makers of these crude implements?

The first great fossil finds pertaining to prehumans were made a little more than a hundred years ago. (By prehumans, we mean ancestors of the hominid line, namely, humans and their extinct close relatives.) Around the time that Darwin published his treatise on biological evolution, a primitive-looking skull was found in a cave in the Neander Valley in Germany. This skull has a low, sloping forehead, a receding chin, and thick ridges over the eye sockets, though it still displays an overall "manlike" appearance. The German word for valley is *thal*, hence the origin of the name, Neanderthal man, attached to the original owner of this skull. Though a bit odd compared to today's human skulls, there is little doubt of its human origin. However, with only one such skull fossil known to exist at the time, it was easy to classify it as a deformed specimen of modern man, which was exactly the approach taken. Even as recently as a hundred years ago, many scientists who embraced the concept of evolution with regard to plants and nonhuman animals were still apparently unwilling to accept its relevance to humans as well.

Only toward the end of the nineteenth century were more Neanderthal-type fossils found in Germany. Less primitive humanlike skulls were also excavated from numerous localities scattered across France, especially near the village of Cro-Magnon. These latter skulls are representative of what paleoanthropologists call Cro-Magnon man, an entire subspecies of human ancestors. Regardless of the scientific designation, the important point is that many of these odd-looking, though clearly human-related, skulls were found alongside the ancient tools. The proximity clearly implies that toolmaking humans of some sort resided in Europe a hundred thousand years or so ago. Since the Cro-Magnon skulls are less primitive than those of the Neanderthal variety, Neanderthal man is regarded, in some quarters, as the ancestor of Cro-Magnon man. Other anthropologists demur, claiming that the Neanderthals represent a divergent branch of hominids that died out about forty thousand years ago. Whichever it was, and for whatever reason, the Cro-Magnons replaced the Neanderthals some four hundred centuries ago. The critical question then becomes: Who were the ancestors of Neanderthal man?

Rich fossil findings have traced our roots much farther back in time. Near the start of the present century, humanlike skulls and teeth were discovered far from Europe, in an arid riverbed in Java, a large island in Indonesia. These skulls, dated to be nearly a million years old, seem even more primitive than those of either Neanderthal or Cro-Magnon man. Yet the size, shape, and overall features of these Java man skull and tooth fragments still resemble those of today's humans. Furthermore, the hole at the base of the skull, through which the spinal cord passes, is positioned in such a way that these creatures must have stood erect.

Astonishing discoveries scarcely more than a half-century ago, these ancient humanlike fossils predictably drew a great deal of skepticism. It's understandably hard for us to imagine that erect humanesque creatures could have existed anywhere on Earth as long ago as a million years before the present. This is a terribly long time by human standards, equivalent to some forty thousand generations of human life. In fact, a million years is more than a hundred times as long as all of recorded history. Alternatively expressed, more than ninety-nine percent of humankind's history is recorded only by the fossils.

Confirmation of these startling findings followed during the first decades of this century when many similar fossils were excavated at widespread locations throughout the temperate zones of planet Earth. Researchers have now uncovered numerous Java man skulls, as well as bones of Heidelberg man in Germany, Peking man in China, and a variety of other ancient though clearly humanlike fossils in Hungary, France, Spain, and Africa. Most of these fossils are on the order of a half-million years old, though some are closer to a full million years old, and a few might even be slightly older. Significantly, these are not skulls of apes. Nor are they skulls of ape-men. They're skulls of humans — erect men and women who existed an awfully long time ago.

Since humanlike fossils dated to be less than about a million years old have skulls and teeth quite closely resembling those of modern humans, all of them are given the designation *Homo*, a Latin word meaning "man." To distinguish these older fossils from those of twentieth-century bones, a suffix is often added to this designation. For example, Neanderthal man, Cro-Magnon man, and fossils of other humanlike creatures dated to have lived less than a few hundred thousand years ago are collectively given the name *Homo sapiens*, meaning "wise man." This is the same biological species as modern men and women, though some researchers prefer to endow the most recent humans of recorded history (including ourselves) with a special designation — *Homo sapiens sapiens*. Undoubtedly another expression of human vanity, this designation of "very wise man" is a highly debatable label, given the labyrinth of global predicaments we've created for ourselves on twentieth-century planet Earth.

In contrast, Java man and other humanlike fossils between a few hundred thousand and a million years old are collectively referred to by the species name *Homo erectus*, meaning "erect man." They were definitely of human stock, walking erect and exhibiting surprising manual dexterity, but their brain volume wasn't as large and their tool use not as advanced as those of *Homo sapiens.*

The trove of humanlike skulls as old as a million years doesn't really solve the central issue. It merely pushes back in time the central question: Who were the ancestors of *Homo erectus?*

The first clue arrived some fifty years ago, though it wasn't until about a decade ago that a reasonably clear line of ascent emerged. An anthropological expedition in the 1920s uncovered a fossilized skull having simultaneously human and apelike characteristics. After the skull was carefully dug up, dusted off, and reassembled from pieces, analysis showed it to have the following curious blend of ape and human traits: an interior skull volume (brain capacity) larger than an ape's, though smaller than a human's; a jaw larger than a human's, though smaller than an ape's; a forehead resembling an ape's more than a human's; canine teeth more like those of a human than those of an ape; a hole at the base of the skull suggesting that this creature had walked upright, or nearly upright.

Such a mixed bag of bone qualities strongly suggests that this creature belongs someplace near the threshold of humanity. All fossils of this hybrid ape/human kind have subsequently been known by the tongue-twisting Latin name of *Australopithecus africanus,* meaning "southern ape of Africa." Unfortunately, the early findings in the sandy soil of southern Africa could not be dated; sand is not radioactive and tends to shift with time. But this discovery focused modern paleoanthropological research on the African continent, where it's been ever since.

Excavations during the past two decades have revealed numerous additional australopithecine skull and tooth fragments. Some of these findings have been made in the same southern African area where the soil hampers accurate dating. In addition, though, numerous similar fossils have been uncovered all along the East African Rift Valley, a giant crack produced by the disjointed drift of that large continent. For example, about twenty years ago, an *Australopithecus africanus* fossil was discovered protruding from a layer of volcanic ash along a dry river gulch at Olduvai Gorge, Tanzania. The ordered layers of volcanic rock could be dated, and thus the australopithecine fossils could finally be set in time. That date is approximately two million years ago, an age estimate corroborated by recent findings. Clearly these protohumans, or ape-men if you will, inhabited our planet a very long time ago.

The official designation of the two-million-year-old skull remains found at Olduvai Gorge is not without controversy. The discovering Leakey family of Britain and Kenya argue that these are the skulls of a species related to but distinct from the australopithecines. In particular, the co-discovery of very elementary stone tools prompted them to suggest a separate species designation, *Homo habilis*, or "handy man." However, the rock chips the Leakeys consider "tools" are most primitive, making it hard to assess how handy these creatures really were. Many paleoanthropologists maintain that these fossils are merely those of advanced australopithecines and not ones deserving of the humanesque status of *Homo*.

The two-million-year-old prehuman ancestors were no larger than a hundred fifty centimeters tall, weighing about fifty kilograms, or roughly five feet and a hundred pounds. They were surely smarter than any other life forms with which they shared the open savanna away from the forests. Their brain was probably not large enough to have facilitated speech, though these creatures may well have communicated using a repertoire of grunts, groans, arm gestures, and other body movements. The more talented members surely possessed dexterous hands — not as good as ours, but good enough to fashion primitive stone tools — and probably keen eyesight. Whatever their full complement of attributes, these creatures seem to have adapted well to their environment, for this is the key to survival.

More recent fieldwork indicates that several variations of australopithecine creatures may have coexisted throughout Africa several million years ago. Hundreds of australopithecine fossils have now been categorized into at least two distinct species of prehumans. One of these species is characterized by a heavy jaw and large

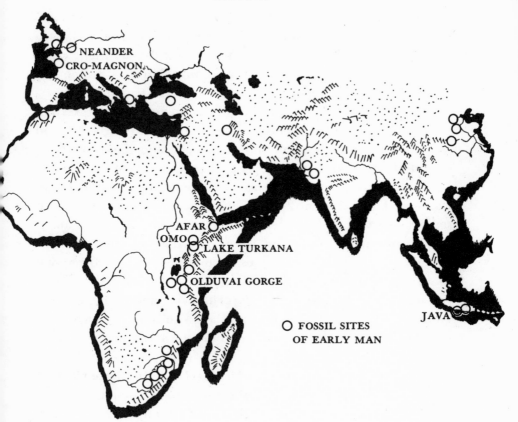

NEANDER
CRO-MAGNON

AFAR
OMO
LAKE TURKANA

OLDUVAI GORGE

JAVA

O FOSSIL SITES
OF EARLY MAN

grinding teeth, suggesting that this species enjoyed a diet of mostly coarse vegetation, much like that eaten by modern gorillas. This more robust type is often called *Australopithecus boisei*, or *A. boisei* or even *A. robustus* for short. The other species, called *Australopithecus africanus* or *A. africanus* or even *A. gracile*, is of the originally discovered southern African variety. This species is characterized by a more slender jaw and smaller molars, suggestive of a more gentle anatomy, and thus of a class of prehumans that probably feasted on meat. These suggestions are just that — suggestions and not conclusions — but they do represent the prevailing view among anthropologists today.

Given these findings, it's natural to wonder if the observed differences in the australopithecine fossils could simply be variations of the same species. After all, today's humans display slight variations — witness the contrast between thin hurdlers and husky weight lifters. This interpretation doesn't seem tenable, however, for all the two-million-year-old fossils of prehuman creatures fall into distinct classes; either skulls and teeth are clearly big and oversized, or they are small and graceful. Could these two types of australopithecine fossils simply correspond to male and female? Again this interpretation seems improbable, because these two classes of fossils are hardly ever found at the same place within sedimentary rock. Unless there was something most peculiar about prehuman cultures that kept tribes of males separated from females, then it would seem impossible that these classes correspond to sexual differences.

Thus, at least two species of prehumans, and quite possibly more, apparently coexisted on Earth a few million years ago. Presumably, only one of these species is our true ancestor.

Expeditions all along the East African Rift Valley have revealed much new information during the past decade. Besides the rich lode at Olduvai Gorge in Tanzania, several groups are helping to spin the thread of our origins by examining fossils found along the shores of Lake Turkana (formerly Lake Rudolf) in Kenya. And, before the (human) guerrilla war in the mid-1970s prevented further digging, particularly well-preserved fossils were found in easily datable volcanic rock at Omo, Ethiopia.

Among the recent discoveries at several of these sites, the most interesting is perhaps of something that's missing: no *A. boisei* fossils are less than a million years old. The fossil record of this most robust prehuman species abruptly ends, implying rapid and unorthodox extinction. The most popular explanation contends that competition between *A. boisei* and *A. africanus* was inevitable. Each biological niche can be filled by only one species, yet here were two prehuman species trying to make a go of it simultaneously. Surprising at first thought, it then becomes clear why the bigger, more robust species was the loser. Despite their larger physique, *A. boisei* found vegetation plentiful, thus producing a rather comfortable way of life. Such comfort, however, is not necessarily conducive to rapid evolution toward an intelligent technological society. The smaller species was almost surely more versatile, quicker, and perhaps smarter. Only basic intelligence could help *A. africanus* capture the less abundant meat needed for survival. As a result, natural selection worked to help generations of *A. africanus* expand their brains, their capabilities, and their niche, all the while apparently crowding *A. boisei* right off the face of the Earth. This theory is supported, not only by the documented demise of *A. boisei*, but also by the recent discovery that only *A. africanus* used primitive stone tools. It's not known if these tools are simply a measure of *A. africanus'* proclivity for manual dexterity and gradual brain development, or if they were actually used as weapons to accelerate *A. boisei's* extinction.

Recognize that, as for many evolutionary scenarios at present, the details have yet to be formalized. Many have yet to be unearthed. Accordingly, several alternate views postulate slight variations in the evolutionary picture sketched here. First there are those scientists who feel that the tool-using creatures of two million years ago should be labeled *H. habilis*, as opposed to an advanced version of *A. africanus*. This mostly amounts to a question of semantics and is the reason why some scientists contend that *Homo* dates back not much more than a million years, while others suggest that some species of *Homo* was around at least two, perhaps three or more, million years ago.

Another view incorporating new discoveries of skull, tooth, and other bone fragments in the Afar lowlands of Ethiopia suggests that humanlike creatures existed nearly four million years ago. Skulls having smaller brains and larger canine teeth than ours have been found near footprints preserved in hardened radioactive ash, suggesting that these truly ancient creatures stood erect. On the basis of these recent findings, some researchers contend that these fossils comprise the best evidence for the creature that must have resided midway between apes and humans — a missing link of sorts. They suggest a whole new species, *A. afarensis,* as the common ancestor of *H. sapiens* as well as the extinct *A. boisei.* Opponents argue that the Ethiopian fossils simply belong to the *A. africanus* species, but acknowledge that that already distant ancestor must be pushed even farther back in time. Still others argue that these fossils constitute evidence for a more primitive version of *H. habilis.* Whichever viewpoint is correct, these oldest humanlike fossils virtually prove that our ancestors walked erect before their brain enlarged appreciably.

So many evolutionary paths are generally consistent with all the fossil data that a cynic might suggest that there are as many possible paths as there are paleoanthropologists. The real problem here is that the present picture of human evolution is based upon just a roomful of partially crushed skulls and broken teeth, most of them found scattered throughout east Africa, Asia, and central Europe. In point of fact, the whole lot thus far uncovered doesn't contain enough parts to reconstruct a single skeleton of an australopithecine; the oldest complete hominid skeleton is that of a sixty-thousand-year-old Neanderthal man.

We're not trying to confuse the issue here. There is an overall evolutionary trend agreed upon by most anthropologists: *Australopithecus* → *Homo,* or near man → true man. The controversies, frequently heated and emotional, essentially concern details — specific dates, emergence of new species in Africa, Asia, or elsewhere, coexistence of various species, invention of new tools, cooperative hunting, language, and other human qualities. They're important details, to be sure. And the emotion is understandable, for these issues involve our own origins. But until many more fossils are unearthed, differing viewpoints will continue to flourish. That's perfectly fine, since each viewpoint is a slightly different hypothesis, and each remains to be tested experimentally by collecting more fossils. This is the way science progresses.

The prevailing view that the *A. africanus* species is the best candidate for the link between contemporary humans and whatever ancestry we share with the apes is certainly instructive but,

even if valid in every respect, it simply pushes the fundamental question still farther back in time: Who were the ancestors of the australopithecines? Here the answer becomes a lot more vague because the fossils are older, more scarce, and less well preserved.

A few discoveries have been made of fossils predating those of the Afar lowlands, while still displaying slight human qualities. For example, bone fossils of an arm and jaw found near Lake Turkana were embedded in rock dated to be approximately five million years old. Most researchers are of the opinion that they belong to the *A. africanus* species, or whatever preceded *A. africanus*, though no one can be sure on the basis of an arm bone and a partially crushed jaw. In addition, a single nine-million-year-old molar tooth has been unearthed at a nearby location. Needless to say, a single tooth cannot be used to trace the ancestry of the australopithecines with any degree of assurance.

Current paleoanthropological thought suggests that creatures having some human qualities resided on Earth more than five million years ago, but it's frustrating that no extensive prehuman fossils have yet been found for the period between four and twelve million years ago. Plenty of fossils from that time span lie nearly everywhere in Earth's soil, but these are usually the remains of animals unrelated to humans.

Tooth and jaw fragments of the oldest creatures having any resemblance to humans or prehumans were discovered in India, after which others of the same age were found in Africa and Europe. Dating of the rocky dirt in which they were buried implies that these fossils are roughly twelve million years old. Despite their age, the jaws in particular still seem to have a mix of apelike and human qualities. The creature's brain capacity and posture are unknown, however, for a complete skull has never been found. There are only a few such fossils, none very well preserved. Yet, most anthropologists contend that this creature, called *Ramapithecus* for the Indian god Rama, may be the ancestor of the australopithecine — a sort of protoaustralopithecine. Again, though, this is only conjecture based upon the meager data currently available. Much more fieldwork and interpretation are needed to create a feasible portrait of our ancestors who roamed Earth some ten to twenty million years ago.

Precisely where and when one species transformed into another cannot really be pinned down much better than the description here, for one life form gradually dissolves into another over the course of history. The fossil record will never document an apelike mother giving birth to a distinctly human infant, or an *A. africanus* mother siring a *Homo erectus* infant. Evolution just doesn't operate that way. Transformations of this sort are decidedly gradual, occurring over long, long periods of time.

All contemporary humans have brain masses of about thirteen hundred grams, weighing nearly three pounds. Variations arise from person to person, though there seem to be no substantial behavioral differences between people with brains of only a thousand grams and those with brains as weighty as two thousand.

On the other hand, most mental patients having reduced cognitive abilities do have substantially smaller brains. Often measuring five hundred grams, the brain capacity of these mentally retarded adults is equivalent to that of a normal one-year-old child. Apparently, the brain mass can be so tiny that its function is considerably impaired, suggesting that a minimum brain mass is necessary for "adequate" human intelligence as we know it. Once this threshold — probably around a thousand grams — is surpassed, normal human behavior is possible.

What about our ancestors? Does the fossil record allow us to estimate the brain size of some of the prehumans that paved the way for our existence? The answer is yes. Paleoanthropologists have been able to sketch a rough outline of the recent evolution of the brain. They do so by estimating cranial volume of the fossilized skulls of our immediate ancestors, assuming that, as is true now for humans, apes, monkeys, and other modern mammals, the brain matter nearly fills the skull.

The partly bipedal and prehuman australopithecines of five million years ago had brain volumes averaging five hundred cubic centimeters. This is comparable in absolute size to the brain of a modern chimpanzee, and is about a third the size of today's average human brain. Thus, fossil evidence supports the hypothesis that our ancestors could walk on two feet before they evolved large brains.

More advanced australopithecines of two or three million years ago (*Homo habilis* to some researchers) had definitely larger brain volumes, most having an average cranial capacity of some seven hundred cubic centimeters. Not only that, but their fossilized skulls have a distinctly different shape from that of their forebears. Developed substantially were the frontal lobe behind the forehead and the temporal lobe above each ear, those brain regions regarded as sites of speech, foresight, curiosity, and, no doubt, many other useful behavioral qualities. Coupled with the fact that these ancestors were fully bipedal, the possibility that they may have fashioned primitive tools implies that at least two significant changes in behavior — toolmaking and bipedalism — were accompanied by equally significant changes in brain volume. Whether this increased behavioral sophistication produced a larger brain or merely resulted from it remains another chicken-or-the-egg conundrum. The fact that bipedalism freed the hands for tasks other than propulsion nonetheless connotes a causal link among upright posture, toolmaking, and brain size.

Fossils suggest that *Homo erectus*, perhaps the first true man and our closest relative, had a brain volume just a bit less than those of our friends and neighbors alive today, on the order of a thousand cubic centimeters. Large and small circular arrangements of stones found alongside the fossilized remains of this species furthermore suggest that our ancestors of a million years ago had domesticated fire and constructed homes outside of caves.

Comparisons of various cranial capacities, then, clearly suggest that the advances made in the last few million years are at least partly related to an increased total brain mass. New behavioral functions, increased neural specialization, and improved cultural adaptations apparently characterized the steady evolution from *Ramapithecus* through *Australopithecus* (perhaps through *Homo habilis*) onward to *Homo erectus* and finally to *Homo sapiens*.

The absolute size of the brain is important, but it cannot be the sole indicator of intelligence. Small-bodied creatures such as spiders obviously have very small brains, whereas large-bodied creatures such as elephants have much larger ones. Yet in many respects spiders act smarter than elephants. This is probably because spiders have a lot less body to monitor and control than elephants. In fact, much of the elephant's large brain is known to consist of motor cortex — enormous numbers of neurons devoted to the process of enabling these huge hulks to put one foot in front of the other without tripping. Hence, it seems reasonable that a better measure of intelligence is the ratio of brain mass to body mass.

Comparison of the brain masses of various animals having similar overall size shows a clear separation of reptiles from mammals. For any given body mass, mammals consistently have a larger brain mass, usually ten to a hundred times as great as those of contemporary reptiles of comparable size. Likewise, the brain sizes of our prehuman ancestors also seem to have been more massive, relative to body mass, than those of all the other mammals.

The creature having the largest brain mass for its body mass is *Homo sapiens*. Dolphins come next, followed by great whales, then apes.

The brain-to-body mass ratio then provides a useful index of the intellectual capacity among different animals. In this way, the fossil record can be used to suggest that the evolution of mammals from reptiles about two hundred million years ago was accompanied by a major increase in relative brain size and intelligence. It furthermore suggests that the emergence of humans some millions of years ago was also accompanied by an additional development of the brain.

∞

More than any other property, the brain most clearly distinguishes humans from other life forms on Earth. The development of speech, the creation of civilization, and the invention of technology are all products of the human brain's sophistication. But what about other forms of life? Are there other creatures now on Earth with comparable sophistication — animals having neural processes enabling them to communicate, make tools, or fashion a society?

Brain-to-body mass ratios suggest that next to humans, dolphins are the smartest animals now on Earth. Dolphins, along with whales and porpoises, are members of a family of mammals whose ancestors were land dwelling. Because of the undoubtedly keen competition among many amphibians several hundred million years ago, the dolphins' ancestors returned to the sea, probably in search of food. No doubt there would have been some disadvantages to such a seemingly backward move, but that ancestral decision probably saved them from extinction.

Dolphins, as we know them today, are well adapted to the sea. Their exceptionally strong bodies are streamlined for deep diving and speedy locomotion. They possess extraordinary hearing, as well as an uncanny sonar system providing them with a kind of underwater vision; this advanced system of echo location, now being studied by human naval officials for military purposes, may employ acoustical radar to map the position and movement of objects in their environment. Dolphins travel in schools or families, have a well-organized social structure, and assist each other when in trouble; females often act as midwife for another dolphin. They're not at all hostile, being anomalously friendly to other dolphins as well as to humans. In fact, dolphins seem to be the exception to the unwritten rule that all friendly species are inherently aggressive as well.

It's not easy to gauge dolphin intelligence. The criteria for human intelligence, controversial in their own right, can hardly be expected to apply to other species. Though the average dolphin brain size is seventeen hundred cubic centimeters, larger than a typical human brain, laboratory tests suggest that dolphin intelligence lies somewhere between that of humans and that of chimpanzees.

In addition to their unparalleled ability to navigate underwater, dolphins are known to communicate with one another by means of a series of whistles, quacks, squeaks, clicks, and other noises that often resemble Bronx cheers. Can we hope to communicate with them someday? Perhaps so, but part of the difficulty stems from the fact that the human range of generating and hearing noise is relatively limited, compared to the much wider auditory range of dolphins. They're known to be able to produce and hear sounds in our communicative range, but to do so requires them to grunt and groan at frequencies lower than normal. Most of the sounds made by dolphins are inaudible to humans, making it improbable that their way of expressing meaning overlaps ours at all. It's not inconceivable that dolphins in captivity have been trying to communicate with us for years. If so, they must be quite discouraged by our lack of response.

The development of interspecies communication will not be easy, whether it is with dolphins, whales, or porpoises. Empirical findings nonetheless suggest that there is some common ground for the future cultivation of dolphin-human links. At least, it seems that both parties are interested in such an enterprise.

Apes are also closely related to humans intellectually and socially. Not only does the fossil record imply this, but the similarities in genetic makeup of apes and humans as well as the behavioral patterns of contemporary apes virtually prove it.

Of all the members of the ape family, chimpanzees have a genetic constitution closest to that of humans; the numbering and ordering of amino acids in average human protein is about ninety-nine percent identical to that in chimpanzees, hemoglobin (blood) in particular being exactly the same. Though gorillas outwardly resemble humans more than do chimps, their genetic structure is a little different, and their daily habits considerably different.

Chimps have the life-style closest to our own. More than any other animal, chimps seem to resemble the ancestor from which other apes as well as humans ascended. By studying modern chimps, behaviorists attempt to discern a little bit of what life was like for our ancestors several million years ago. Indeed, chimps' present attributes, adopted environment, and social behavior can tell us something about the evolutionary events that led to the emergence of humans, and especially the rise of intelligence.

Earlier in this century, several attempts were made to study the life-styles of caged chimps in zoos. But it soon became clear that the intricacies of ape society could be unraveled only in the wild. And "the wild" means just that; chimps and their ape relatives often live in remote places, as might be expected, for their niche must differ from that of humans, lest all of us be unable to coexist. Most chimps are shy and totally unaccustomed to being watched by intruding humans. Many inhabit inaccessible mountains, while others stay up in the treetops of thick jungles. Reaching the appropriate places often proves tricky for scientists, as do the problems of which chimp characteristics should be studied, and how to interpret the data once collected.

Organized fieldwork during the past few decades has enabled researchers to conclude that chimps and other semibipedal apes are clearly more intelligent than monkeys and other quadrupedal animals. The erect posture offered by bipedalism frees the hands, and the resulting manual dexterity in turn provides a wealth of new opportunities for living. For example, the uncanny handiwork of modern chimps has been observed as they routinely fashion an implement by stripping leaves from a tree branch, after which they insert it into a hole in a termite mound, remove it slowly, and carefully lick off the termites clinging to the branch.

Which was the cause and which the effect — ability to stand on hind legs or capacity to manipulate with the hands — is currently a matter of dispute. Instead of the above possibility, namely, that erect posture caused toolmaking, the situation may have been reversed: The need to manipulate food may have helped humans' ancestors to become permanently upright over the course of millions of years. It may turn out that these two critically important evolutionary traits are so hopelessly intertwined as to preclude a determination of which was the original instigator. Actually, each trait may have contributed to the other in a complex sort of way: primitive bipedalism may have led to a small degree of manual dexterity, then increased hand use accelerated the transformation toward more erect posture, which in turn fostered the development of even more sophisticated tools, and so on. This is an example of positive feedback, whereby the development of one attribute stimulates another, causing further and faster mutual development. Such feedback reinforcement was probably a key mechanism throughout many phases of prehuman evolution.

The physiology and habits of modern chimps imply many things

about our australopithecine forebears. Chimps are small enough to get around in the trees, yet large enough to ward off most predators while on the ground. They can be particularly formidable when traveling as part of a large group, as they often do.

The favorite food of chimpanzees is fruit, especially ripe figs, though they also eat meat and birds' eggs, as well as small snakes, lizards, and insects. They are known to experiment with different foods, revealing some innate curiosity.

Basic chimp intelligence manifests itself in many ways. They have an open, free behavior enabling them to try new things. In addition to using twigs as tools, chimps have been observed to use rocks to smash objects, to wave large branches overhead to intimidate enemies, and to employ grass as a sponge to hold water.

Perhaps even more interesting than their expressions of curiosity is the recent report that chimps may show some degree of self-awareness. For example, when exposed to mirrors, chimps rapidly progress from treating the image as if it were another chimp to recognizing it as themselves. Once thought to be inherent only to humans, self-recognition seems to be part of the intellectual repertoire of these remarkable apes.

Chimps are also known to be observant copycats. The young learn readily from their elders, as well as from their human trainers. Though their lack of vocal equipment prohibits chimps from speaking, a few of them are now able to communicate symbolically with humans by means of sign language routinely employed by deaf people. Most significantly, some chimps have used such symbolic gestures to communicate with other chimps. Chimp-to-chimp conversation of this type suggests that chimps' intelligence incorporates substantial communicative and learning ability, implying that they're really a lot smarter than any scientist would have argued a decade ago. Parrots and seals can also imitate, but there's a difference here: Other animals can be trained to imitate, whereas chimps seem to have a childlike ability to learn by imitation.

Chimps are furthermore sociable, though in a highly stratified way. All groups of chimps show a clear hierarchy, comparable in many respects to twentieth-century human assemblages, whether in the military, business, academic, industrial, or political sector. One or a few males usually dominate a host of subservient chimps, thus ensuring some degree of stability rather than the constant infighting that might otherwise engage a completely free society. The social hierarchy does not, however, seem to stifle their curiosity or amicability. Some chimps are forthrightly altruistic, sharing food with other members of their group. They are thus not entirely self-centered, and show some consideration for others, even including chimps who are not their immediate offspring.

Chimpanzee society is so complex that it takes about fifteen years for a newborn chimp to reach maturity. Like human adolescents, young chimps apparently need many years to learn everything necessary to integrate themselves into their social organization. In a certain sense, young chimps are schooled by their parents. But because they learn so well, it's impossible to tell how much innate intelligence they really have.

That chimps really learn in their formative years demonstrates that their environment plays a large role — at least among modern chimps. Other completely unrelated species, such as bees and ants, also have organized societies, but they don't really learn. Environment seems to have little bearing on insect knowledge. Laboratory studies show them lacking freedom of individual expression while tackling their daily business; insects demonstrate little, if any, of the curiosity needed to try new things. Consequently, while insect society is quite clearly organized, insect behavior is not all that complex. The social organizations of insects are almost entirely programmed by genetics.

The extent to which intelligence is genetically preprogrammed, as opposed to being environmentally endowed, is most controversial. The gene (nature) versus the environment (nurture) debate has forged a whole new interdisciplinary field of research. Called sociobiology, it's the study of the biological basis of social behavior. It attempts to determine the social instincts within any community of life forms by drawing together the basic principles of psychology, genetics, ecology, and several other seemingly diverse disciplines. The goal of sociobiology is to identify the inheritable traits that mold societies.

Sociobiology is an expansion of the study of classic biological evolution to include society. Another name for it might be social evolution. In social evolutionary terms, the fitness of an individual being is measured, not just by its own success and survival, but by the contributions that he or she makes to the success of his or her relatives, namely, those who share some of his or her genes. These contributions are often self-sacrificing ones, and thus can be classified under the general heading of altruism — unselfish devotion to the welfare of others — a fancy word for love. Whereas the catchphrase for classic biological evolution is "survival of the fittest *individual*," that for sociobiological evolution would be something like "preservation of an entire *society*."

Many human societies seem to be founded on altruistic behavior. Extensive studies have virtually proved this to be so for ants and bees, as well as some other animals. For instance, wild dogs regularly regurgitate meals in order to feed their young; some species of birds postpone mating to help rear their siblings; "soldier" termites explode themselves, spraying poison over armies of ants, whenever a termite colony is attacked. This behavior is always the same, regardless of where and when the dogs, birds, or termites happen to act. They perform like programmed machines. Behavior so rigid and uniform has prompted many researchers to suggest that it's probably determined genetically. If so, then each trait, act, or duty has its own gene, which is inherited in much the same way as body size, shape, and structure. The principal role of these behavioral genes is to preserve the species.

It would appear that, while imprisoned within the bodies of life forms, the genes control all. In the extreme, life forms may exist for the purpose of perpetuating the genes — the selfish and unaltruistic genes.

Sociobiology is currently controversial largely because its proponents argue that its central tenets can be extended from insect societies to the societies of higher life forms, including humans. Problems always arise when scientists make grand pronouncements about our own species. The trouble stems from the fact that some aspects of human nature are often not what we think proper. The main issue is this: To what extent does human behavior mimic insect behavior and its largely genetic basis? In other words, which has the dominant influence over the actions of humans, nature or nurture? This is the root of the controversy — human understanding of human affairs.

Researchers generally fall into two groups, each granting that environmental influences play the major role in human behavior. One group maintains that environment is the only important influence; behavioral differences among different humans are exclusively due to social, cultural, and political factors. The other group contends that the gene has considerable importance. It may weigh only one-tenth in the balance against the environment, but enough that many traits (aggression, envy, sympathy, love, fear, intelligence, among others) are mostly predestined in humans. This second group suggests that little can be done to alter basic human behavior because, for the most part, it's biologically dictated by the genes. This group presumes that, for example, the behavior of women who go hungry to feed their children, or the behavior of persons who risk their lives to save a drowning swimmer, is not governed completely by free will. Instead, paralleling an insect's desire to preserve its own species, such behavior is an unconscious reaction built into the genes to ensure survival of our own kind.

Should this second group be correct, it will be important for psychologists and psychiatrists to pay close attention; the way people act may be, in large measure, biologically predetermined. Sociologists should also take note, for sociobiology may someday give them quantitative methods by which to test their overabundance of unsupported statements. Indeed, economics, law, and politics might eventually become part of the newly emerging subject of sociobiology.

Altruism, love, and curiosity are attributes associated, not only with humans, but also with chimps and probably other animals as well. This is not to suggest that chimps are nice and gentle all the time. They resemble humans in yet another way, namely, their occasional desire to exert unnecessary aggression. Some hostility within and among species is a normal, perhaps even essential, ingredient of biological and cultural evolution. Without some aggression in the guise of competition, few if any species could survive or adapt to a changing environment. But unnecessary aggression is another thing entirely.

Recent field studies in Tanzania show that some chimps occasionally murder other chimps for no apparent survival-related reason. Premeditated, gangland-style attacks were directed by a large group of male chimps on a smaller group of males and females that had previously broken away from the larger group. Over the course of five years, each member of the splinter group was systematically and brutally beaten. All died. Only young males initiated the attacks, which occurred only when the victims were isolated from the others. Hands, feet, and teeth were often used by the attackers, though occasionally field-workers noticed stones being deliberately thrown. The hope, of course, is that comparative studies like these will uncover the reasons behind not only chimp misdemeanors but human belligerence as well, perhaps helping to guide the future survival of the human species, which, it would seem, can no longer tolerate intraspecies aggression.

Despite present controversies about many details, behavioral studies of modern chimps have contributed immeasurably toward a consensus concerning the ascent of humans: As apelike animals resembling chimps began leaving the forests for the savanna some tens of millions of years ago, they were probably forced to become more sociable in order to survive. The origins of our social organization may well have been shaped by this new, harsher environment in the open plains where there would have been less food, reduced protection, and thus a greater need for group cooperation. These hardships nonetheless gave our ancestors a chance to learn. Their experience accordingly grew over the course of millions of years. The parochial mentality of forest-living animals was replaced by the wider perspective of our savanna-dwelling ancestors. This suddenly larger world effected the evolution of larger brains capable of storing a dramatic increase of raw information.

Migration from the trees to the plains was a renaissance of sorts. Once it commenced, the race was on — a race to inhabit entirely new niches, to develop whole new ways of life, and eventually to become intelligent in the technological sense.

∞

Nearly all researchers agree that our ancestors must have survived by hunting and gathering food for the bulk of the past few million years. The acquired traits of pursuing and eating meat were probably exported from the forest to the savanna, whereupon they were enhanced if only because of the relative lack of fruit in the open plains. Though most inhabitants of modern civilization no longer regard themselves as hunter-gatherers, this was indeed the job description of all our ancestors from several million years ago until the rise of agriculture some ten thousand years ago.

How do we know that early humans, even the advanced australopithecines, hunted? The evidence is twofold, and both parts are found in the fossil record. First, at numerous dwelling sites throughout the East African Rift Valley, scattered bones of a variety of large animals are often found near those of our ancestors. These animal bones do not constitute intact skeletons, but rather strewn debris suggestive more of a picnic than a natural death. Second, and more convincing, tools made from stones are often found alongside the remains of two-million-year-old australopithecines as well as of all the more recent human species. These stone artifacts have endured for millions of years, and it seems safe to conclude that even earlier ancestors might have brandished wooden tools that did not endure.

From the shapes of the stone tools unearthed at Olduvai Gorge, it appears that many of these egg-sized implements were used to chop, cut, and prepare food for easy consumption. Many others, though, suggest use as weapons. This is especially true of the rounded stones probably used to maim or kill when thrown. Other stones clearly resemble club heads, and they were probably used for exactly that — hunting by killing with a club of some sort. As noted earlier, these were perhaps the "tools" used by the advanced *A. africanus* (or *H. habilis*) to exterminate its relative, *A. boisei*, about a million years ago.

Stones were not used only for tools and weapons. They also provided the basis of early homes. A two-million-year-old locality in Olduvai Gorge, for example, contains a circular stone structure theorized to have been the foundation of a hut of some sort. This kind of primal stonework predated what is now popularly referred to as the Stone Age.

Spanning a period from roughly a million years to approximately ten thousand years ago, the extent of the Stone Age depends upon the place excavated. Whatever the dating, this new age is distinguished by increasingly intricate stone tools, including various types of hand axes, cleavers, spatulas, and scrapers. A steady transition from rather crude tools to more advanced ones can be clearly noted alongside the fossil record of biological species. Hence, the beginnings of the Stone Age are customarily associated with the onset of *Homo*. Pre–Stone Age toolmaking doubtless accelerated the evolution of the first true human beings.

Much of this development of stone implements preceded the enlargement of the brain. The earliest stone-wielding creatures had brains just a little larger than those of modern chimps, not much more voluminous than five hundred cubic centimeters. Tool use and bipedal posture, hallmarks of manual dexterity, were powerful evolutionary developments, fundamental changes that precipitated whole new opportunities for living. Unbeknownst to our ancestors of the time, their tool-y chips of rock were the beginnings of a manufacturing society, a technological culture. The difference between stony spoons and jumbo jets is only a matter of degree.

The threshold of technology, then, is hard to pinpoint exactly, but it seems to have occurred more than a million years ago. The beginnings of cultural, as opposed to utilitarian, activities may be nearly as old, for brightly colored mineral pigments have been found alongside skeletons of the earliest of the true humans. Even the advanced australopithecines may have had some use for ritual, as geometrically arranged pebbles are often found alongside their remains.

Only toward the end of the Stone Age do more industrious and sophisticated undertakings become evident: Technological advances such as the construction of the wheel some fifty thousand years ago, and the invention of the bow and arrow about ten thousand years ago, were matched by cultural advances such as the oldest deliberate burials in certain European and Asian caves nearly seventy thousand years ago, and the beginnings of prehistoric art on the cave walls of western Europe about thirty thousand years ago. These were uniquely human inventions — cultural products of *Homo sapiens*, including Neanderthal and especially Cro-Magnon man.

The precise path of human evolution during the past million years is tricky to follow in detail. Whatever truly made us human involved the creative products of emotion and imagination — qualities virtually impossible to define scientifically. The causes of recent evolution include not only biological factors but cultural and technological ones as well. A jumbled feedback system has come into play among the increase in brain volume, the innovation of culture, the invention of improved technical skills, and the development of complex verbal communication and social organization. Changes were very slow at first, but they have markedly accelerated within the past hundred thousand years or so. Whatever the reasons, these many innovations have enabled *Homo* to enjoy unprecedented success as a life form on planet Earth, for we alone can technologically address the fundamental questions.

Changes that effected humanity were evolutionary, not revolutionary. They occurred by means of the usual adaptations to a gradually changing environment. But they happened in an increasingly rapid way. Broadened opportunities for living some million years ago quickened the pace of evolution. And it hasn't slowed since.

Environmental changes act as the motor of evolution, allowing some life forms to adapt successfully while forcing others to become extinct. Apart from seasonal changes taking place from month to month, and continental drifts occurring over millions of years, planet Earth apparently experiences global climatic changes enduring for thousands of years. Climatic records dating back at least a half-million years have been derived by a variety of techniques, including analysis of core samples taken from the icy polar regions and of sandy sediments extracted from below the sea floor, as well as land-based geological evidence for freeze-thaw cycles.

Climatic changes are among the most important environmental alterations on Earth. As shown by the data, our planet has cycled through numerous episodes of cool, dry climate — intermittent periods popularly known as ice ages. Though the data are sketchy,

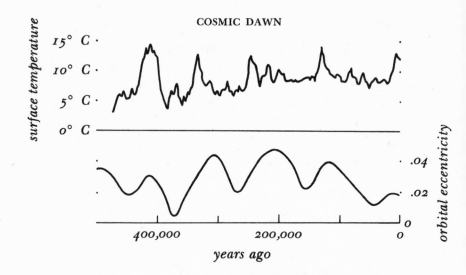

each cold ice age, as well as its opposite warm interglacial period, apparently lasted several tens of thousands of years. We now reside in an interglacial period — a temporary thaw of sorts, a geologic lull before we head back into the deep freeze.

What causes these cycles of heating and cooling on our planet? Some scientists contend that glaciation increases during periods of global volcanic activity when ejected dust reduces the amount of sunlight penetrating Earth's atmosphere. Others suggest that the periodic alterations in Earth's magnetic field cause the protective Van Allen belt to disappear, thereby sporadically allowing unusually high doses of solar radiation to heat the ground and thus decrease glaciation. Still other researchers have offered the possibility that ice ages could be triggered on our planet by variations in the output of the Sun itself, passage of Earth through an interstellar dust cloud, deep circulation of Earth's oceans, or any one of a long list of other proposals.

Recently, however, scientists have found convincing evidence to support yet another theory, which argues that subtle though regular changes in Earth's orbit around the Sun initiate the ice ages. These changes are the combined result of three astronomical effects, each caused by the normal gravitational influence of other planets on the Earth: first, alteration in the *shape* or eccentricity of Earth's elliptical orbit around the Sun; second, periodic changes

236

in the *tilt* or wobble of Earth's axis; and third, *precession* of Earth's orbit caused by the variable position of the other planets, each of which gravitationally pulls. Sometimes the combination of these three effects results in abnormal heating (such as we are now experiencing); at other times, solar heating is distinctly reduced, producing widespread glaciation and a decline in global temperature.

This hypothesis of an astronomically induced ice age is currently favored among the majority of scientists, primarily because samples of sea-floor sediments show that, over the course of the past half-million years, tiny sea creatures known as plankton have thrived at certain times, while barely surviving at others. Studies of the abundance of fossilized plankton known to prefer warm or cold water provide estimates of the prevailing temperature during their lives. This inferred sea temperature correlates well with the expected heating and cooling of Earth by means of the combined astronomical effects just noted. It would seem, then, that slight peculiarities in Earth's orbital geometry are primarily responsible for triggering ice ages; further research will tell if they are the only trigger.

Regardless of their cause, ice ages must have had a profound impact on the evolution of life on Earth. The most recent major glaciation began nearly a hundred thousand years ago, and the climate had pretty much returned to the way we now know it by ten thousand years ago. At the peak of this ice age, some thirty thousand years ago, an ice sheet a kilometer (nearly a mile) thick extended from the North Pole far enough to cover much of the northern United States as well as a good deal of north and central Eurasia. Earth's overall surface temperature averaged several degrees Celsius lower than it does today, while the sea level, with much water locked up in icy bergs, was about a hundred meters (three hundred feet) below current values.

The last ten thousand years have seen the glaciers retreat, the coastal plains flood, the vegetation climb toward northerly latitudes, and the ocean and atmosphere warm. The transformation in climate from the peak of the most recent ice age to its present state occurred rapidly, by geological standards. These widespread

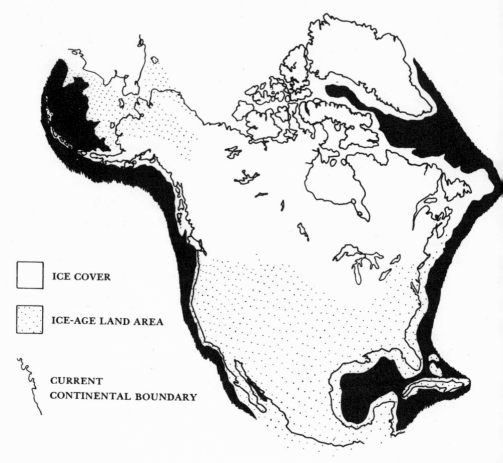

ICE COVER

ICE-AGE LAND AREA

CURRENT
CONTINENTAL BOUNDARY

changes in Earth's environment surely helped unleash the many advances made by our human ancestors during this period of rich innovation. Humans were forced to adapt — to adapt biologically, culturally, and rapidly — to changes throughout the air, sea, and land. The motor of evolution had quickened.

In addition to stimulating the rapid emergence of innovative products, the ice ages themselves may have accelerated the migration to and colonization of new lands. For example, scientists know that humans didn't evolve in the Americas; there's no fossil evidence for apelike creatures from which they could have ascended. The anthropological consensus contends that humans arrived in the New World quite recently — a few tens of thousands of years ago. Precisely how they did it is unknown, but during one of the more recent glaciations, enough water would have been wrapped up in ice to have permitted humans to walk dry-shod across the Bering Strait between what are now called Alaska and Russia. This is in fact the prevailing view; native North, Central, and South Americans are descendants of Asians who wandered here only a few hundred centuries ago. Once settled in the Americas, these migrants developed arts, language, products, and numerous other cultural amenities. Whether civilizations of the Americas experienced cultural evolution totally independent of those in Eurasia, or whether they had some as yet undiscovered contact with them, is a central point of controversy in modern archeology.

Several factors helped make us effective humans. The discovery of fire's usefulness is among the foremost of them. Humans have been using fire for light and heat for nearly a million years; archeological scrutiny of caves in France has revealed fossil hearths at least that old. The essence of fire, therefore, was basically appreciated a very long time ago, and was undoubtedly used to provide warmth in colder climates. But it seems that its broad usefulness went unrecognized until more recently. The practice of cooking food, for example, was probably developed only a hundred thousand years ago. Techniques needed to fire-harden spears and anneal cutting stones are likely to be still newer inventions. Significantly, the widespread use of fire was one of the last great steps in the domestication of humans.

Beginning about ten thousand years ago, humans devised ways to extract iron from ore and convert silica into glass, as well as to cast copper and harden pottery. Many other industrial uses were realized for fire, including the fabrication of new tools and the construction of clay homes. Unfortunately, though, there's no

agreement about the ordering of many of these fundamental discoveries. Just how one invention paved the way for another is so
far unclear. For instance, some archeologists argue that, after heat
and light, baking clay to fashion pots was the oldest organized use
of fire — even predating the regular cooking of food. Others, however, maintain that the need for pottery arose because cooking was
already established; indeed, the earliest uses for pottery must have
been for cooking and storing food.

When inventions are intimately linked in this manner, which
was the cause and which the effect is never quite clear. Both the
motivation for and the exact time of an invention are difficult to
establish. Some of the more crucial sequences may never be pinned
down. Of one thing we can be sure: Scores of new mechanical and
chemical uses of fire were mastered during the past ten thousand
years, and a few may have been discovered even well before that.

The ancient clay, metal, and glass products can still be found
in the bazaars and workshops of Afghanistan, China, Iran, Thailand, Turkey, and other generally Asian countries. More modern
results of these early technologies are evident in the cities of steel,
concrete, and plastic surrounding many of us in the twentieth
century. Changes of this sort are not without problems, however;
the basic need for fire has often been accompanied by grim aftereffects, not the least of which include environmental pollution and
energy shortages only now becoming evident.

Development of language, that marriage of speech and cognitive
abilities, was another crucial factor in making us cultured. The
onset of meaningful communication is probably related to hunting and toolmaking by means of a complex feedback mechanism
that led to the further development of each. Symbolic communication, such as body movements and arm signals, would have been
an obvious advantage in big-game hunting. Likewise, communication of some sort must have been needed to convey such
rudimentary skills as simple tool use and toolmaking. Equally
important, language ensured that experience, stored in the brain
as memory, could be passed down from one generation to another.

Though there's no direct evidence, some type of primitive language could have been employed a million or more years ago. More sophisticated language came much later, though again it's hard to pinpoint its origin and evolution with any degree of accuracy. Artifacts made by early humans provide some clues to the cognitive abilities and manual skills required to develop such advanced techniques of communication as speaking and writing.

Excavations of caves, mostly in Europe, have uncovered a wealth of small statues and bones having distinctive markings or etchings. The oldest such artifacts date back about fifty thousand years. Markings on a few of them imply that the statuettes had symbolic functions of some kind. Since Neanderthal man's larynx couldn't possibly have uttered all the sounds of modern speech, the etchings on these objects are thought by many researchers to represent a primitive kind of communication. Recent analyses have reinforced this view by finding the same repertoire of distinctive markings repeated on many of the statues and bones. Apparently, the markings are neither accidental nor simply decorative; they're symbolic. These engraved artifacts are among the earliest known attempts to better the grunts and groans used not only now by modern chimps but also long ago by our ancestral australopithecines.

Credit for being the first to write texts usually goes to the Sumerians, who resided on a flat plain bordering the Persian Gulf in what was then called Mesopotamia. Nearly six thousand years ago, this ancient civilization had developed an intricate system of numerals, pictures, and abstract symbols. Thousands of Sumerian clay tablets have now been unearthed, and each shows that a stylus of wood or bone was used to inscribe a variety of characters. Estimates suggest that the Sumerians' basic vocabulary was large — no fewer than fifteen hundred separate signs. Wild animals such as the fox and wolf, as well as technological aids such as the chariot and the sledge, are clearly depicted, but Sumerian texts remain largely undeciphered. Because they contain more than just pictures, the messages on these tablets represent an advanced stage in the evolution of the art of writing.

The Sumerian writings have survived primarily because they

were inscribed on baked clay. If earlier civilizations were responsible for originating many of the symbols, they probably wrote exclusively on papyrus or wood that would have decayed long ago.

The prevailing view among anthropologists is that writing evolved from the concrete to the abstract. Somewhere during the period from several tens of thousands of years ago to several thousand years ago, the pictures and etchings on the bones and cave walls must have become increasingly schematic, utilizing a single symbol to represent an entire idea. This evolution could have been deliberate, or it could have been the result of carelessness on the part of ancient scribes. Either way, it suggests that pictures were the precursors of symbolic writing, which in turn led to the character writing of modern times, as in this book you're now reading.

Statue and bone markings just mentioned may have had something to do with the origin of ancient science. These etchings of several tens of thousands of years ago seem to correlate with the periodic lunar cycle to depict phases of the Moon — perhaps the first attempts to keep track of the seasons. Engravings of this sort, as well as large murals on the walls of caves, suggest that men and women of the late Stone Age were conscious of the periodic seasonal variations in plants and animals. While some archeologists prefer to interpret these deliberate bone markings as a simple arithmetical game, the above hypothesis contends that these artifacts were among the very first calendars, or scientific timekeeping instruments.

The earliest writings clearly describing use of scientific instruments were not set down until many thousands of years later. Three-thousand-year-old hieroglyphics partially document the Egyptians' knowledge of the sundial, but the oldest sundial excavated to date is a Greco-Roman piece of stone built about two thousand years ago. The Greek and Roman sundials reveal important refinements over what we know of the earlier Egyptian models, telling not only the hour of the day, but also the day of the year. Neither the length of an "hour" nor the construction of the calendar were the same as they are now, but these early scientific instruments were nonetheless able to predict the orderliness of daily, seasonal, and yearly changes on Earth, thus identifying the basic rhythm of the farming cycle.

Megalithic monuments, the most famous of which include the pyramids of Egypt and Stonehenge in Britain, exemplify early human knowledge and use of the sky. Stonehenge, for example, can be used to predict the first day of summer, and possibly eclipses as well, by the alignment of big stones with the rising and setting of certain celestial objects. Several other structures like it, though perhaps not as grand, are scattered across Europe, Asia, and the Americas. In Central America, ancient "temple" pyramids have miniature portholes through which celestial objects, particularly Venus, can be viewed at propitious times of the year. Astronomers of the Middle-Age Mayan civilization were essentially priests, granted the power to use their knowledge of celestial objects to determine the destinies of individuals, cities, and even whole nations.

By a thousand years ago, these Central American cultures had influenced many tribes of North American Indians who roamed the plains of what is now called the western United States and Canada. Recent discoveries of numerous "medicine-wheel" structures made of boulders arranged in various patterns of rings and spokes clearly demonstrate the cross-cultural fertilization. Though the precise use of these stone configurations is unclear, as most Indians had no written language, some researchers suggest that these structures may have been used as calendars to mark the rising of the Sun at certain times of the year.

Thus, although modern scientific research, including instruments like the immensely useful microscope and telescope, dates no farther back than five hundred years, we can be sure that pre-Renaissance people were adept in elementary astronomy, mathematics, mechanical engineering, and numerous other technical endeavors. Indeed, the roots of technology go far back. Our ancestors seem to have been a good deal more technically sophisticated than many researchers have cared to acknowledge.

* * *

Mention of the Mayan astronomer-priests of ancient Mexico brings to mind another factor that contributed to the cultural evolution of humans — a more lofty factor that drives us to know who we are and to understand how we fit into the general scheme of things. Indeed, the epitome of culture is the search for truth, the need to know ourselves and the world around us. The desire and ability to undertake this search truly identify us as humans, distinguishing us from all other known life forms. To be sure, understanding is the major goal of science, though science shares this goal with other disciplines. Religion, the arts, philosophy, among others, represent alternative efforts to appreciate who we are, and how we came to be.

For thousands of years, humans have recognized that the best way to dispel mystery is to understand it. The twenty-thousand-year-old cave paintings of southern France may be the oldest traces of early magico-religious ceremonies in Earth's dim recesses. The cave-wall images seem to depict rites in which elaborate myths perhaps linked the hunting men and the animals they killed.

Reliance on the supernatural and the use of mystical activities are more clearly documented near the very beginning of civilized history. Sumerian inscriptions of five thousand years ago record myths explaining how gods had created humans to be their slaves. This system guaranteed that food, clothing, and other necessities of life would be provided to priestly households or temples in order to please the gods, or at least appease them. Such a society divided managers from the managed, priests from the plebeians. Apparently, anyone professing knowledge of even the simplest celestial events was able to subjugate the masses; in the eyes of field workers, anyone who foreknew the seasons, for instance, must have had a special relationship with the gods, and therefore deserved to be obeyed. No longer required to spend time producing their own food, the priest-masters of ancient Sumer were able to develop skills and knowledge far greater than humans had ever before attained.

Such mythmaking, perpetrated on the populace, forced further specialization, while permitting legions of humans to labor on

244

social and technical tasks that we today identify as vast irrigation projects, and monumental templelike structures rising high above the Mesopotamian plain.

Sumerian poetry of several thousand years ago clearly documents how Sumer religion incorporated a systematic theology of human, worldly, and cosmic phenomena. Aspects of nature — Sun, Moon, storms, thunder, and so on — were personified, with humanesque beings playing roles of gods in a divine political society, the whole of which was ruled by the resident god of the sky.

Such a set of beliefs proved powerful, for the system was complex, and the populace largely uninquiring. For several thousand years, the priests of Mesopotamia bamboozled the public with increasingly intricate speculations. Even the surrounding barbarians, the ancestors of the ancient Greeks, Romans, Celts, Germans, and Slavs, were convinced that the gods of Sumer ruled the world. Apparently, myths become truths if upheld long enough.

In some ways, a stratified social system helped to maintain cohesion and uniformity; priestly leadership offered a stabilizing influence, as do all religions in principle. But Sumerian inscriptions also tell of quarrels, prompted by water-rights disputes, and often fostered by rival religious factions and coalitions, that gradually became life-or-death struggles. By a few thousand years ago, dozens of Mesopotamian cities were armed to the hilt, with military organizations rivaling priestly governments. It can be surmised that the concept of kingship originated when Sumerian priests decreed that the gods needed a representative among men — a strong-armed chief priest of sorts to adjudicate quarrels or to crush the opposition. It wasn't long thereafter when clustered societies, united under the influence of one king or one god, began demanding adherence to this, that, or another thing, all with the objective of telling our ancestors who they were, and how they fit into the Universal picture.

Today's plethora of differing religions and philosophies testifies to the fact that theological and cognitive ideas are not subject to experimentation, and thus will never be universally acceptable. They offer stabilization, perhaps, but how can these ideologies fail to destabilize as well, especially in light of conflicting sects and viewpoints in our modern world? The upshot is that belief alone does not make the unknown known.

Ultimate reliance on the authority of the experimental test is the one feature that clearly distinguishes the scientific enterprise from all other ways of explaining nature. Initiated in ancient Greece, and used sparingly through the centuries, most notably by Aristotle and Saint Augustine, the scientific method rapidly developed during Renaissance times to become one of the primary criteria in the search for truth in the physical Universe. It is *the* primary criterion used to generate the late twentieth century view of cosmic evolution — not a tradition of beliefs told to people, but a set of discoveries offered to them.

The epitome of culture may well be the ability to seek the truth about ourselves and our Universe, but an even more basic factor of associated import concerns the ability we've developed to *desire* to seek the truth. Just what is it that allows us, even drives us, to ask the fundamental questions, and to attempt to find solutions to them? The answer is consciousness, that part of human nature that permits us to wonder, to introspect, to abstract, to explain — the ability to step back, perceive the big picture, and examine how our existence relates to the existence of all things.

How did consciousness originate? When did humans become aware of themselves? Is consciousness a natural consequence of neurological evolution? Some researchers think so, but they can't yet prove it. Others demur, suggesting that some specific, perhaps unlikely mechanism is required for its development, for they argue that the capacity for imagery and imagination represents something more than just a continued accumulation of neurons.

Records of ancient history are sketchy, making difficult any attempt to document the onset of self-awareness. Some researchers contend that consciousness, as we know it, is not manifest in ancestral records until about three thousand years ago. This is about the time when some of the writings in ancient texts take on an abstract or reflective tone. Thus, there's little doubt that people wondered about themselves several thousand years ago, but it's unclear if this was the first time they began to do so. If consciousness did originate that late in time, then we must be prepared to assume that cultures can become highly refined without developing personal consciousness. Our historic ancestors would have had to invent just about every cultural amenity except consciousness, and to have lived until quite recently in a dreamlike, essentially unconscious state.

Other researchers argue that human consciousness developed long ago. If modern chimpanzees, which display rudimentary self-awareness, do, in fact, mimic our australopithecine forebears, then consciousness could have been a factor millions of years ago. That it evolved several tens of thousands of years ago, about the time of the invention of the bow and arrow, is another popular assertion. Indeed, the development of the long-distance weapon could possibly be viewed as that giant step finally granting humans the freedom to innovate, to begin to evolve culturally, to wonder.

An obvious reconciliation of these seemingly divergent views suggests that prehistoric humans did develop a crude sense of consciousness a million or so years ago, but only recently did they become sophisticated enough to reveal that sense of wonder and self-awareness in their writings. In the absence of a good experimental test, however, we are left wondering how we learned to wonder.

The most recent one percent of human history — the past ten thousand years — has clearly seen major and rapid cultural innovations. The glaciers had retreated, creating warmer and wetter environments, thus allowing the land to flourish. Our hunter-gatherer forebears had spread to occupy, though sparsely, every portion of the habitable globe except the Arctic and Antarctic. And they were fashioning implements and ideas to enhance their survival. But of all the factors that contributed to the rapid rise of modern humans, the invention of agriculture was surely among the most important. Tilling the land made available a reliable source of food to feed the swelling numbers of people on Earth.

Hunter-gatherers rather rapidly transformed into agrarian-culturists, beginning with the domestication of plants and animals some ten thousand years ago. Archeological data show clear evidence for whole new methods of subsistence by eight thousand years ago. Systematic crop planting and livestock raising near stable village settlements permitted large increases in the population. Not only did more people survive, but many more migrated to colonize every nook and cranny on the planet. The basic techniques of agriculture spread like a weed on the wind from Asian and Aegean localities, and especially from the area now called Greece. Hunter-gatherers everywhere gave way to farmers and herders. Change was rampant. Urbanization and ultimately industrialization were not far behind.

The environment continued to change, though not as slowly as nature would have had it. After all, humans themselves had become a factor. They were less at the mercy of the environment than conversely, for men and women had gained some mastery over matter. Our ancestors had become agents of change.

Civilization had moved into high gear. It had taken an awfully long time after life originated, but highly organized and manipulative life forms had finally arrived. One thing led to another; lifestyles multiplied. Aided by irrigation systems built alongside river valleys, the art of farming developed dramatically. The human population multiplied rapidly, especially in urban areas along waterways such as the Nile River in what is now Egypt, and the Tigris and Euphrates rivers running through what is now Turkey, Syria, and Iraq. Specialized crafts were refined to serve the populace of these growing communities: Metalwork, fine ceramics, shipbuilding, and woodcuts all show up clearly in the archeological record beginning about six thousand years ago. The record is best documented for southwestern Asia (the Middle East), though surely pragmatic and artistic progress ensued at other geographical centers as well.

Though much of this preceded recorded history, the consensus has it that urban societies and ripened economies, as well as complex social and political systems, were the rule, not the exception. Agriculture, industry, and commerce were fully established many thousand years ago. And they have persisted to this, the twentieth century.

Be sure to place the later developments of civilized humans into perspective. Thousands, tens of thousands, even millions of years tend to merge into a temporal blur after a while. To appreciate the time frame for the more familiar advances of recorded history, consider the following analogy.

Imagine the entire lifetime of the Earth to be fifty years, instead of five billion, thus making each megacentury a "year." This time scale may then be compared to a human life span, allowing salient features of Earth's history to become more comprehensible. Within the realm of this analogy, we can say that no record whatever exists for nearly the first decade. Rocks hardened relatively quickly thereafter. Life originated at least thirty-five years ago, when Earth was no more than fifteen years old in our analogy. The planet's middle age is largely a mystery, although we can be reasonably sure that life continued to evolve, and that continental collisions continued to build mountain chains and oceanic trenches.

It wasn't until about six years ago, in our fifty-year analogy, that abundant life flourished throughout Earth's oceans. The same life did not come ashore until about four years ago. Flowering plants and primitive animals prevailed across the surface not more than two years ago. Dinosaurs reached their peak about a year ago, only to disappear suddenly eight months ago. Manlike apes became apelike men only last week, and the latest major ice age occurred only yesterday. *Homo sapiens* did not appear until several hours ago. In fact, the invention of agriculture is only about one hour old, all of recorded history a half hour, and the Renaissance a mere three minutes in the past.

In short, you would need a microscope to see the highlights of recorded history on the scale spread across the endpapers of this book. Yet it is within that microscopic duration that we humans have richly probed, reasoned, and discovered much about ourselves and our Universal environment.

∞

Throughout the past million years or so, biological and cultural evolution have been inextricably interwoven. Their interrelationship is natural, for the development of culture bears heavily upon the environment, a prime factor affecting the course of biological evolution. Cultural innovations enabled our immediate ancestors to circumvent some environmental limitations: Cooking allowed them to adopt a diet quite different from that of the australopithecines, while clothes and housing permitted colonization of both drier and colder regions of Earth.

Likewise, present cultural creativity enables us — twentieth-century *Homo sapiens* — to challenge the environment. Advanced technology provides for us a means of flying in the atmosphere, exploring the oceans, even journeying far from our home planet. Change now quickens, and with it the pace of life. Culture, it would seem, is a catalyst, speeding the course of evolution toward an uncertain future.

If there's any one trend that has characterized the evolution of culture, it's probably an increasing ability to extract energy from nature. Over the course of the past ten thousand years, humans have steadily mastered wheels, agriculture, metallurgy, machines, electricity, and nuclear power. Soon, solar power will emerge in its turn. Each of these innovations has channeled greater amounts of energy into culture. Indeed, an ability to harness larger energy sources is the hallmark of modern society. It's the way we can, in principle, continue to avoid chaos and to order life on planet Earth, thus locally circumventing the fundamental laws of thermodynamics. But, in practice, it's also a cause of many of the sociopolitical complexities in which we, twentieth-century humans, now find ourselves embedded.

EPOCH SEVEN

FUTURE

Rest or Progress?

billions of years ago

HUMANS ARE NOW the preeminent intelligence on planet Earth. We're the only species able both to communicate culturally and to construct technologically. We're the only ones capable of knowing our past and worrying about our future. Just how wise we are, however, is an issue of considerable uncertainty.

Where do we go from here? What is our future? Though these are not easy questions, we do know one thing: Our Sun is destined to run out of fuel, balloon into a red-giant star, and engulf several of its planets, perhaps even Earth. That will doubtless be the end of civilization in our Solar System.

On the other hand, the Sun will not so perish for another five billion years, a time so remote as to be nearly incomprehensible. That leaves plenty of opportunity for life on Earth, should it endure, to undertake galactic engineering projects and other grand ventures literally out of this world.

What about shorter time scales, say a million, a thousand, or even a few hundred years in the future? Is there any way at all to predict further evolution of *Homo sapiens* while extending our trek along the cosmic arrow of time?

Surely, we would be conceited, vainglorious, and downright pretentious to regard ourselves as the final product of Universal change, the very pinnacle of cosmic evolution. Change has accompanied time's passage since the very start of the Universe. Just because technologically intelligent life has achieved dominance on a single planet, there's no reason to think that change will now cease.

Change must furthermore persist for all tomorrows if we're to survive as a civilization. Change or perish. That is the vital code for the continued viability of all matter, including life.

The future is a tricky subject. To comment on it is to run the risk of saying nothing concrete. The future is especially troublesome to predict when life is involved. As a case in point, predicting the destiny of our civilization is more difficult than predicting the destiny of the Universe. It may sound ludicrous that we can know more about the future of the Universe in bulk than about the future of life on our own planet. But life weighs heavily on our civilization, while hardly at all on the whole Universe; and while the Universe obeys the laws of physics, civilizations legislate their own laws.

The fate of the Universe depends on only one thing — its mass density, a term that scientists can define and whose quantity they are now trying to estimate. The destiny of our civilization also seems to depend heavily on a single term. But that term is *humanity*, a very general expression seemingly impossible to quantify. Even humanity's nature is hard to grasp fully. Webster defines it as the "quality of being humane; the kind feelings, dispositions, and sympathies of humans."

Here's another way of perceiving the riddle of the future. As mentioned previously, the business of a physicist is to comprehend nature well enough to be able to predict the response of matter to a variety of circumstances. The route of a baseball moving through air, for example, is now precisely understood. Knowing the mass of the ball, the mass of the Earth, the gravitational forces between them, the air resistance, the ball's momentum and spin, and a few other physical factors affecting the ball, scientists can predict with great accuracy the future trajectory of this piece of matter through space. To predict the "trajectory" of *life* through *time*, however, is a much tougher puzzle. There are just too many nonphysical factors involved: individual and group sociology, national and international politics, biological and cultural behavior, and a host of other unquantifiable parameters all of which will influence the future of civilization.

Claiming knowledge of the pathways along which cosmic evolution will proceed henceforth is probably equivalent to dabbling in science fiction. Nonetheless, it is possible to examine certain boundary conditions that will affect the future of life on Earth. These boundary conditions represent hazards of a global nature — environmental factors, political decisions, economic sanctions, technological aftereffects, and a plethora of other ailments destined to have an adverse impact on the future of our ever-shrinking world.

Here are four examples of boundaries that we must circumvent in order to survive as a civilization:

Overpopulation, along with its attendant plights of food and energy shortages, is sure to have a negative effect if it continues even at reduced growth rates. This is an example of a problem caused by the actions of many people, and one that could gradually increase in severity.

Self-destruction could result in severe suppression or even extermination of life on Earth. This type of problem could conceivably result from actions of only a few people and, in the form of nuclear holocaust, for instance, could befall us nearly instantaneously.

Genetic degeneration is a representative boundary condition that could eventually intensify many of the detrimental qualities of life on Earth. Once thought to belong to science fiction or at least to the distant future, the potentiality of this problem is real and could well begin to show its ugly consequences sooner than we think.

Finally, silicon-based circuits — computers — are increasing their smarts at an unexpectedly rapid pace. Though no one can say for sure, computers might someday threaten to subjugate us, perhaps even sending us down the path of extinction already trod by ninety-nine percent of Earth's life forms.

These boundary conditions are unlikely to be peculiar to planet Earth. Some, perhaps all, of these problems will be encountered by all emergent civilizations elsewhere in the Universe — should there be extraterrestrial intelligent life anywhere else in the Universe.

Without touching upon our prospects for ultimate survival (this depends largely on the true nature of humanity), we note the implications of each of these potential crises, together with some changes seemingly required to overcome them.

∞

Overpopulation is a tough issue for many people to appreciate, especially for those of us living in the "developed" countries. Everyone recognizes that there are numerous inhabitants on Earth, nearly five billion at present. Surely, that's a large number, but the planet is also a large place. So what's the problem? The problem is that the world population is not stable. Like everything else, population changes, and it's currently changing toward increasing numbers a lot faster than most people realize or than Earth can tolerate.

Since it's impossible for a human even to count to one billion in a lifetime, there's no easy way to appreciate the full magnitude of such a throng of humans, let alone the multitude of other life forms on planet Earth.

Still, some awareness of the population problem can be gained by considering the population density. Since we've specified the density of virtually everything else in the Universe, why not compute the density of human life as well? Here, the population density is defined as the total number of people inhabiting a given area of space.

In Australia and Canada, where there remain vast tracts of uninhabited land, an average of about five individuals occupy every square mile. In the Soviet Union, the largest nation on Earth, the density is about thirty people per square mile. The United States and Europe average nearly sixty and two hundred people per square mile, respectively. And in some Asian countries, India and Japan, for example, the population density soars to about five hundred (India) and eight hundred people per square mile. (Halve these numbers to convert to people per square kilometer.)

But these are average numbers and, as such, not terribly informative. Why? Because people aren't spread evenly across the globe. Like all types of matter, life tends to cluster. On Earth, three-quarters of human life is coagulated in only two percent of the land. There, in cities, the population density is much greater. For example, in Boston proper, the five boroughs of New York City, and Manhattan at noon, there are fourteen thousand, twenty-seven thousand, and one hundred thousand people per square mile, respectively.

These numbers impart some feeling for the concentration of the human species on Earth today. In and of themselves, they don't represent a real threat; Earth can support us. The problem, as noted earlier, is that the world's population is not nearly stabilized.

Each year, we currently produce a net increase of almost eighty million people. This estimate takes into account both the current birth- and death rates, and is no small number. It's equivalent to a third of the entire population of the United States being added to Earth each year.

A total world population of nearly five billion, and a yearly addition of eighty million newcomers, implies an annual growth rate of almost two percent. The fact that the rate of growth has decreased in recent years in the United States, Canada, and several Western European countries has little effect on the overpopulation problem; North America and Western Europe combined house hardly more than fifteen percent of the world's population. The growth rate remains high in the Southern Hemisphere.

What are the general implications for continued growth of the world's population? Examined broadly, there are a number of theoretical scenarios.

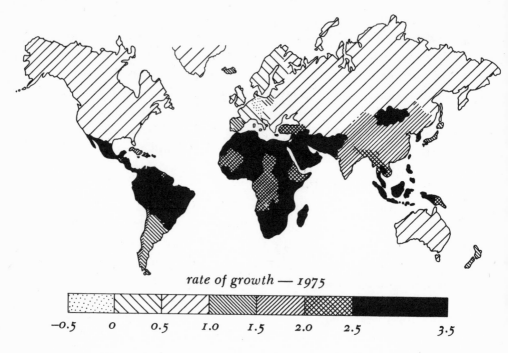

rate of growth — 1975

−0.5 0 0.5 1.0 1.5 2.0 2.5 3.5

If the annual growth rate remained constant, the world population would increase uniformly, producing a linear rise in population so that, for example, five centuries hence the world population would have reached about forty billion persons. This tenfold increase in population would cause current population densities to increase by ten, with average separations among neighbors decreasing accordingly. Five hundred years sounds like a long time, and indeed it is much greater than a typical human lifetime. But it's a mere wink of an eye in the cosmic scheme of things.

It's important to note that constant growth doesn't mean a constant population, a common misconception. It means just what is stated: constant *growth*. Only zero growth guarantees a completely stabilized world population, where the number of births equals the number of deaths.

Population increase quicker than linear growth is also possible. For example, an exponential increase would start out slowly,

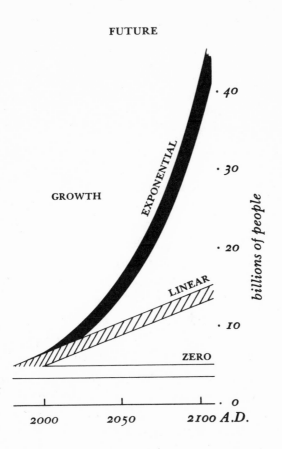

FUTURE

GROWTH

EXPONENTIAL

LINEAR

ZERO

billions of people

· 40

· 30

· 20

· 10

· 0

2000   2050   2100 A.D.

would eventually overtake linear growth, and would then rise substantially faster than any linear increase.

Exponential growth can be alternatively labeled explosive growth. It sneaks up, increasing gradually for a long time, then surges while numbers increase dramatically over a short duration. In the case of population, development of this sort would mean a slow and steady rise in the numbers of world inhabitants over the course of thousands of years, followed by a sudden and phenomenal increase over the course of centuries or even decades — a veritable population explosion. For example, two percent exponential growth predicts a population of more than a hundred trillion after five centuries, a great deal more than that foreseen by linear growth analysis.

257

These, then, are three general possibilities for the increase of anything, population or otherwise: zero, linear, or exponential growth. Still, actual demographic data seem to obey none of these possibilities. Examination of the real situation on Earth shows the population to have increased at a rate even faster than exponential growth!

A big-picture view of census data for the past several thousand years compels us to conclude that humans are now within the domain of catastrophic population increase. (Christian persecution and the black plague represent the only marked departures in world population growth during all of recorded history.) The culprit is again the annual growth rate. When considered over thousands of years, the annual growth rate has neither remained constant nor increased linearly; the *growth rate itself* has risen dramatically during the past century or so. For example, in the past several thousand years, the population has doubled many times, though each time it doubled a lot more quickly. The population roughly doubled to two hundred million people from 5000 B.C. to 1 A.D., a time period spanning fifty centuries. The next doubling time then shortened considerably to fourteen centuries. By 1800 A.D., only four centuries later, the population had doubled once more. Succeeding doubling times have continued to decrease, reaching a hundred years by the turn of the present century, sixty years by mid-twentieth century, and a current value of about forty years, or less than half a century. Such dramatic decreases in the population doubling times indicate nothing less than truly explosive growth.

Another way to visualize population growth is to note that, of all the humans who ever lived on this planet, five percent are now alive. Within two decades, this percentage will have doubled.

Given the finite size of planet Earth, there's no escaping the fact that this extraordinarily rapid population growth is unhealthy for life here. No stretch of the imagination can visualize anything short of disaster, should this proliferation of humans continue unchecked. But will it continue? Many people routinely answer no,

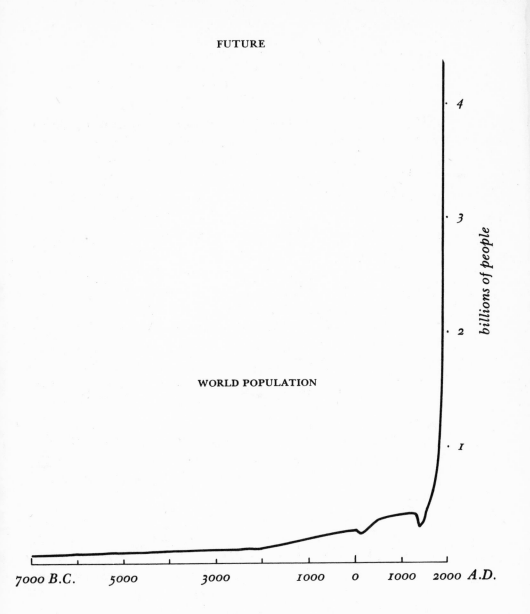

WORLD POPULATION

billions of people

· 4

· 3

· 2

· 1

7000 B.C.   5000   3000   1000   0   1000   2000 A.D.

it can't. Yet what makes them smugly sure that this problem will not become absolutely unmanageable?

Two points are worth noting. First, some demographers maintain that the prospects for overpopulation are not as grim as presented here. They claim that the annual growth rate has now peaked, and that the problem is under control. Some indicators do in fact suggest that the worldwide growth rate is now on the decline. It's hard to know for sure, since census figures are often inaccurate for many of the heavily populated Asian and Latin American countries.

Current population statistics often seem riddled with confusion. For instance, substantial declines in the rate of growth have been reported in the last year or so for Costa Rica, Sri Lanka, South Korea, Fiji, Indonesia, Panama, Colombia, the Dominican Republic, Malaysia, and Thailand. Many people think this means that the population problem is behind us, despite the fact that all these countries house hardly more than five percent of the planet's people. In fact, most news releases of this late-1970s United Nations study ran under the headline "World Population Decline Documented." This is nonsense. Population has not decreased in the world. Nor has it decreased in these countries. It hasn't even stabilized. The world population is still increasing, and rapidly at that.

A decreased growth *rate* still means growth; it only postpones problems. For example, an annual growth rate of one-and-a-half percent predicts a doubling of the population every forty-seven years, a mere dozen years more than for the two-percent rate considered earlier. Even a growth rate of only two-thirds of one percent would cause the world population to double every century. These time scales are not long in the cosmic scheme of things. To avert an overpopulation problem, the growth rate must be decreased substantially and permanently.

A second point: It's easy to be fooled by slick presentations of demographic data, or of any statistical data, for that matter. When census figures are examined over the course of only a few years or

billions of people

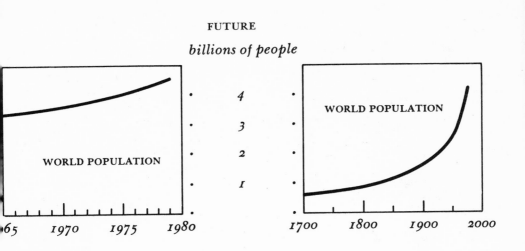

even a few decades, the population problem seems to disappear. The "little picture," sometimes known as the "politician's plot" because of the one-term foresightedness of nearly all elected officials, shows a steady but not drastic rise in world population. The very same function, when plotted over the course of several centuries, however, shows the full magnitude of the problem. This "big picture" gives a completely different impression, forecasting a population that would skyrocket to unbelievable numbers only a few centuries hence.

Always seek the big picture. Don't let statisticians proffer the trees for the forest. Small-term trends offer little information, especially in the cosmic scheme of things. Regardless of the quantity considered, always request the full history of its change over as much time as possible. There's nothing to lose and possibly lots to gain by examining the comprehensive, long-term view. It's the only way to envisage the panoramic drift of matter through time.

The implications for continued growth at exponential or faster-than-exponential rates are simple. All accounts point toward catastrophe. Only madmen and some economists think otherwise. Prolonged growth along the "explosive" portion of the steeply rising curve — that part of the curve at which we now find ourselves — predicts that world population would theoretically approach infinity inside a thousand years.

261

Does an infinite population make sense? Of course not. The population could never become infinite. Planet Earth is of finite size, and it's impossible even to visualize an infinite world population for the same reason that it's impossible to appreciate infinite densities within a black hole or an infinite volume for an open Universe. Infinities are nonattainable.

When speaking of physical quantities that theoretically approach infinity, scientists really mean that we are currently incapable of explaining the properties characterizing infinite density, infinite volume, or infinite whatever. When describing infinitely small or large values, the laws of physics, as we now know them, break down.

In the case of population growth, something is indeed going to break down, but it's not going to be the laws of physics. Long before the world population even approaches infinity, it's more likely that humanity itself will crumble.

The upshot is that a genuine decrease in world population growth is now required if our civilization is to avert a monumental problem not too far in the future. There's little doubt that such a change will occur; the concern is how it will occur. We shall either achieve the change with Machiavellian effort, or it will happen to us with Malthusian suffering.

Emphasis should be on *change* and *now*. Change is required, and it's required now. Because of our position on the explosive part of the population curve, the present time is pivotal. Our generation — not the next one — must effect planned change, lest it come by means of war, pestilence, and famine. Such change can be best achieved by viewing the big picture, recognizing the problem, and then altering the philosophical outlooks of governments, religions, society.

It's natural to wonder if emigration to space could help alleviate the world's population problem. If overpopulation is a result of Earth's limited size, then why on Earth stay here? After all, Renaissance seafarers relieved European overcrowding by discovering the

Americas, Australia, and other new lands. Now that the entire planet seems headed toward a worldwide state of crowding, why can't modern spacefarers just track down some stars having nice new planetary abodes for Earthlings to emigrate to and settle?

Though advocated by some, this approach is nonsense. Interstellar expansion is acutely more difficult than most people realize. Scientists are presently unaware of any other star that definitely has planets. Even if we did know of such a star system, and even if that system were among the closest to us, it would still be far too distant to explore with current skills. The thought of trucking hordes of people there to relieve crowding on Earth is another problem altogether. Interstellar emigration is surely out of the question now, and all analyses of futuristic spaceflight techniques based on the known laws of physics suggest that it will probably never become feasible.

In the absence of many readily available planets, some researchers have urged the construction of large space colonies. Their on-paper designs have accorded such colonies a wide variety of geometrical configurations, featuring tubes, rings, spheres, and sundry polyhedrons. All in all, these colonies essentially amount to large steel and glass enclosures usually spanning two to thirty kilometers (or one to twenty miles). Set into spinning motion, these glorified tin cans experience centrifugal force much like that discussed earlier in the case of protogalaxies and protostars. Because the inhabitants are envisioned as residing on the *inside* walls of such colonies, this outward push would naturally pin them to the walls, thus simulating gravity, and enabling them to overcome the weightlessness normally encountered in outer space.

Various proposals now on the drawing boards envision upwards of hundreds of thousands, even millions, of people inhabiting a single colony. There they would reside for their entire lives in a controlled environment that would ostensibly provide for their every need.

Astonishing numbers of people, including a surprising number of scientists, have recently become enamored of the idea of assembling and colonizing numerous space stations in the neighborhood of Earth and beyond. Yet there would seem to be significant problems of an engineering, sociological, and even philosophical nature.

Construction alone would be overwhelmingly difficult. Aluminum and silicon ore for a colony's metal and glass structure could conceivably be mined from the Moon, as has been proposed. It's unclear how much raw material would be required to fabricate even one colony, especially in view of the structural integrity needed to maintain an airtight environment and to protect the inhabitants from lethal cosmic radiation. Many engineers suggest that nearly a billion tons (a trillion kilograms) of materials are needed for a typical colony, a greater tonnage than that of all the ships of all the world's combined navies. If they are right, then given the abundance of aluminum, silicon and oxygen interlaced with lunar soil, the assemblage of raw materials for just one such colony would require strip mining much of the Moon's surface to a depth of nearly a meter. Space colonization would be, to put it mildly, an absolutely colossal engineering task.

Nor would life in a space colony, once built, be a picnic. Some designs stipulate so many people inhabiting a typical colony that the population density might exceed that of Hong Kong. Even with lesser densities, who knows what might be the psychological and physiological consequences for people cooped up in a rotating space colony for their entire lives. Seasoned travelers are occasionally bothered by being confined on commercial aircraft for only a few hours. Even trained professionals of naval submarine fleets can find it troublesome being isolated for more than a few weeks in an artificially enclosed environment that provides for almost every need. A seventy-year hitch in a spinning space colony would seem to require one helluva lot of adaptation.

The need for structural rigidity leads to additional sociological problems, absolute security being foremost among them. Aside from external threats arising from sabotage or intercolonial conflict, ever-present internal dangers might make space colonies completely vulnerable. For example, the torsional stress on the exterior walls of many colonies is predicted to be near the structural limit — so great, in fact, that any projectile could conceivably crack a glass wall upon even slight impact, potentially unzipping it along the full length of the entire colony, thus allowing the pressurized environment to escape. We might regard the result as a genuine population explosion, or at least a rapid evacuation of the populace! Accordingly, projectiles of all sorts would seemingly have to be banned in space colonies. No baseball and not much other activity could be tolerated, and certainly no rebellions. No guns or bombs either, not a bad idea in and of itself. But to ensure the absolute elimination of bombs, which could render a colony helpless, space societies would be forced to ban the practice of chemistry. And to prohibit chemistry implies banning books. Opponents view space colonies as "Huxley hives," "Bradbury boxes," "Orwellian outhouses," or just plain totalitarian tubes.

As blueprinted now in their early design stages, space colonies are, depending upon the beholder, novel life vehicles or luxurious death traps. Is there a compromise position between these two extreme attitudes? Maybe not, for space colonies present a philosophical dilemma as well.

Space colonies demand a completely untried venture in living: They require us to put something where there is nothing. They are designed to bottle a breathable and warm environment, with people living on the inside of a fully enclosed structure. It's not the colony itself that's foreign to space as much as the attempt to create a permanent environment where there naturally exists no such thing. Outer space is the closest known thing to a perfect vacuum, though it harbors sparse but lethal radiation. Any attempt to create a permanent habitat where there is essentially nothing at all is wholly novel and fundamentally unique to any of our living routines on Earth. Our air is not tightly bottled, our environment not artificially warmed. We inhabit the outside of a rock whose environment is kept naturally intact by nature's gravity. Planets are not without problems, but their environments are based on *natural* foundations — solid bases provided by nature. In fact, planets may be the only objects capable of supporting permanent, safe, and livable environments.

These criticisms do not extend to all types of construction in outer space. Space is perfectly good for flimsy structures — giant antennas for efficient communications, solar power collectors to energize our civilization, astronomical observatories to improve our perception of the Universe. Suspended in the weak-gravity environment of space, such man-made structures require little rigidity, since there are no net forces aside from the steady breeze of particles emanating from the Sun. Literally fabricated from aluminum foil, such tenuous structures would be relatively cheap, expendable, and, most of all, realistic.

We explore the oceans, but we don't live in them. We research the polar regions, but no permanent, self-sustaining commune is lodged there. We travel in the upper atmosphere, but no one inhabits it. Let's explore space and travel through it, but let's postpone, perhaps permanently, attempts to colonize it.

One thing is certain: Space colonies cannot possibly solve the population glut or any other problem now facing our civilization. Though some proponents of space colonization actually argue that interstellar emigration could alleviate the problem of Earth's growing population, a simple calculation proves this viewpoint foolish. The current annual increase of nearly eighty million people can also be expressed as a daily increase of some two hundred thousand people — the projected capacity of typical space colonies now on the drawing boards. Just to neutralize the current growth of world population would require, not only building one of these giant habitats every day, but also arranging for the daily launch of a couple of hundred thousand people from the surface of Earth.

It's of little help if our civilization someday in the future miraculously gains the advanced technology to construct fleets of space colonies overnight. To make any impact on world population, such colonies would have to be built and ready for occupancy before we reach the steeply rising portion of the population curve. We don't have space colonies now, and we need them now. It seems unlikely that the required technology and resources will ever be secured to export even a fraction of the world's population.

Overcrowding is unlikely to be a problem peculiar to intelligent life on Earth. The underlying reasons for a burgeoning population seem to be Universal, suggesting that any intelligent life forms on any other planet would experience a similar predicament at our level of development. Biologically speaking, no life forms are likely to increase their intelligence faster than their number, thus leaving all of them unprepared to emigrate to outer space when the overcrowding begins. How can we be sure of this without knowing anything about the sociology of galactic aliens? The reason is that the basic cause of a population explosion has little to do with sociology.

Overcrowding results from the way matter is structured, or, more precisely, by the sequence of discoveries through which any intelligent life form is bound to unravel that structure. Dramatic growth in population results largely from the discovery and harnessing of bacteria — in Earth's case, through the suppression of disease. Ability to construct large space colonies and especially to travel to distant star systems, on the other hand, depends on the discovery and utilization of atomic nuclei. Since bacteria on Earth are invariably larger than atomic nuclei, and since we can presume matter to be structured similarly elsewhere, cells will always be discovered before atoms, and hence the invention of medicine will always precede that of nuclear technology. Thus, intelligent life forms on any planet seem destined to encounter overcrowding before gaining the ability to solve it via interstellar emigration.

∞

Overcrowding is a problem, not only because it leaves less room for everyone on Earth, but also because it taxes our planet's supply of nonrenewable and even renewable resources. It spawns a host of other burdens, the need for food foremost among them. Indeed, some researchers argue that food shortage is the natural way to check population growth. Citing biological studies, they note that a limit on the supply of nutrients causes the reproductive growth of microorganisms to stabilize after a brisk period of exponential increase. True enough, but is limiting the food supply the intelligent way to go about it? Malnutrition and even cannibalism might work for bacteria in a petri dish, but they are unacceptable solutions to the problem of human crowding. Theoretically at least, humans should have come far enough toward intelligence to better nature's solution of death by starvation or predation. In practice, we must do better than bacteria, for countries that have both hunger and nuclear weaponry aren't going to remain idle. Let's hope that those who advocate famine or warfare as a natural solution to the overpopulation predicament are not thereby demonstrating a measure of that elusive quality, humanity.

The severity of today's food shortage can be appreciated by noting that nearly three-quarters of Earth's population currently inhabits underdeveloped countries where substantial pockets of periodic famine occur regularly. Despite starvation in an appalling number of localities on our globe, the present food production of all the world's countries is actually sufficient to feed everyone now on the planet. The problem at hand results from the fact that about twenty percent of all processed food is lost because of poor storage or sheer wastage.

Present famine could thus be alleviated at once by efficient distribution of foodstuffs around the world. But global famine is not now nearly the problem it will soon become, for it seems unlikely that food production will be able to keep pace with the rapidly growing population.

Most people capable of reading this book ignore the food problem, first because they themselves are well fed, and second by claiming that synthetic food will rescue us from worldwide starvation. But there's no getting around the fact that some people are starving now. We don't have synthetic food now, and we need it now. Furthermore, to manufacture artificial food sometime in the future will inevitably require more energy, the other fundamental problem arising from our burgeoning population.

* * *

Energy is the common denominator for all technological societies. Energy is required to operate automobiles, trains, aircraft, and other machines that aid movement on our planet; to enjoy telephones, radios, and televisions that permit us to supplement face-to-face exchanges; to fabricate clothing and houses that augment our body's thermostatic mechanism and that enable us to reside in terrestrial (and someday extraterrestrial) sites normally unsuited for humans; to practice medicine and nutrition that make possible longer and healthier lives; to create books and computers that help us remember all that we know. All of industrial production, not only the synthesis of foodstuffs but also the extraction of resources and the manufacture of daily goods, requires the use of energy. Most human activity has come to rely on it.

It would seem that a central predicament now confronting us is that there's simply not enough energy to go around. But that's only a superficial concern, expressed by selfish societies that happen to be alive today, and that primarily worry about filling their automobile gas tanks tomorrow. When the big energy picture is examined, we recognize that the real problem is just the opposite: Our civilization may soon be producing too much energy.

Although less than ten percent of the world's estimated oil capacity is gone, the current rate of oil usage will ensure depletion of the remaining supplies in less than forty years. Within little more than a generation, then, our planet will be mostly oilless — for all practical purposes devoid of a rich resource that is essentially unrenewable. Over the course of about a hundred years, our civilization will have thoroughly exhausted a fossil fuel that took hundreds of millions of years to stockpile.

This is one of the legacies we are destined to leave to posterity. Looking back at us historically, our great-grandchildren and all those who succeed them will recognize that it was those twentieth-century humans who gobbled up all the oil reserves nature provided our planet. Indeed, the large view of world oil consumption resembles a thin flame in a long, dark night.

What are we to do, then? How can we fuel our technological civilization? Some propose nuclear fission, the current tech-

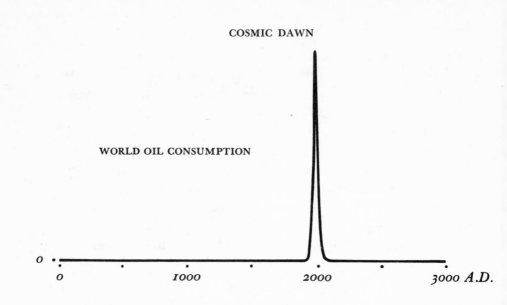

WORLD OIL CONSUMPTION

0 $\cdot$ 0   *1000*   *2000*   *3000 A.D.*

nique used in nuclear power plants to produce energy. But fission also produces radioactive wastes highly toxic to our environment and possibly to our genes, though just right for thermonuclear weaponry.

Some propose nuclear fusion, the very same process that yields energy in stars, and which leaves far less radioactive disorder. But fusion isn't yet mastered, and it's likely to be many decades or longer before it's ready. '

Some propose coal, another nonrenewable resource of planet Earth. But potentially grave pollution problems accompany the overt burning of coal; what we burn up as coal will almost surely come back down as acid rain. Furthermore, most of the world's coal supplies lie in America's Midwest, where the land will have to be stripped naked. Proven reserves will last a few centuries, but after that, they too will be gone. And then what?

Even advocates of that newly discovered resource, conservation, recognize their suggestion to be only a temporary stopgap. Surely, meaningful conservation of unrenewable resources offers the best attack on the world crunch to be experienced by all nations throughout the remainder of the twentieth century. However, conservation does not address the crux of the long-term energy issue.

Last, some suggest that all these energy schemes utilizing non-renewable resources are ridiculous. Indeed, it would seem rather foolish for any intelligent civilization to ransack its own planet for sources of energy, when unlimited amounts of it can be captured from its parent star — in our case, the Sun. Though solar power is economically infeasible at present, its development does seem to be the wisest direction in which to invest capital funds. Let's face it, if dumb plants can harness solar energy, then intelligent animals ought to be able to do as well.

Current shortfalls of energy aside, many people lack the fore-sight required to appreciate the guts of the energy dilemma. The real problem before us is not which of these energy alternatives to embrace as our civilization moves toward the future. The funda-mental problem here concerns our incessant increase in the pro-duction of energy from any source and by any technique. Why? Because energy is heat.

Heat is an unavoidable by-product in the extraction of energy from wood, coal, oil, gas, wind, atoms, the Sun, and *any* other source. Regardless of the basis of energy, Earth is constantly sub-jected to heat generated by our industrial civilization. We already experience it in the big cities that are warmer than their suburbs, and near nuclear reactors that warm their nearby waterways. While, admittedly, this heat is currently an imperceptible burden on the environment, it's now on the rise, obeying a classic expo-nential curve like that for population, and thus destined to become troublesome almost after it's too late to do anything about it. Sig-nificant heating will disrupt the delicate balance between energy arriving from the Sun, and that reradiated by the Earth, possibly destroying the natural thermal equilibrium that keeps our planet reasonably comfortable. Though few people appreciate the fact, we are polluting the air with heat.

Several estimates of planetary heating suggest that, should Earth's average surface temperature increase by as much as a couple of degrees Celsius (four degrees Fahrenheit), then severe environmental consequences can be expected. Melting of the polar ice caps is the foremost concern. The Arctic region is nothing more than a large iceberg, the disintegration of which would not raise the world's water level, just as the melting of floating ice cubes doesn't affect the liquid level in a glass. But decay of the southern Antarctic region, a huge ice-covered landmass that isn't afloat, could raise the sea level by as much as seventy meters (two hundred feet), inundating coastal cities that are abodes for large segments of the world's population. We're not talking about flooding beach resorts and marshlands, but major population centers built up over centuries because of a reliance on sea and river trade. How many of the world's great cities can you name that are not perched on the banks of waterways?

Not only would oceanic waters rise, but atmospheric water vapor would increase as well. This translates as swollen cloud cover, which would allow the long-wavelength solar radio radiation to reach Earth's surface, but not permit a good deal of the re-emitted, shorter-wave infrared radiation to escape into space. Nor would much of the industrially produced infrared radiation (heat) be able to penetrate the clouds. The trapped radiation would bounce around in the atmosphere, heating it even more, thereby causing more melting, more flooding, more atmospheric water, and still more cloud cover. If Earth were to heat sufficiently to initiate ice-cap melting, nothing short of a technological miracle could halt a whole chain of environmental effects from running out of control until our entire planet was shrouded in clouds, seethingly hot, and definitely not an abode for life. Such a greenhouse process warms the interior of glass-enclosed hothouses during winter, and the insides of automobiles on hot days; the same process operates full blast on cloud-cloaked Venus, heating its surface to a hellish temperature sufficient to melt lead.

Could the excesses of technological civilization really trigger such dire ramifications? How much energy can all our technological devices — automobiles, stoves, factories, whatever — produce before Earth's surface temperature increases to the brink of starting this runaway process? With the total incidence of solar radiation on planet Earth known, a simple calculation shows that our average surface temperature would increase by a couple of degrees Celsius whenever the total power emitted from within reached a few thousand trillion watts. Since the power now used by all Earth's inhabitants totals nearly ten trillion watts, a few hundred times as much energy production, by any technique, even including solar methods, could be expected to trigger a melting of the polar ice caps. That's only eight doubling times. Given the fact that energy consumption is currently increasing exponentially at an annual rate of several percent, a straightforward calculation suggests that within only two centuries industrial production will have approached the level of threatening to barbecue life on Earth.

This heating of the biosphere on planet Earth will be further accelerated if large amounts of coal are burned, as is now being advocated by mostly nonscientific bureaucrats. The trouble with coal is that it pollutes the air with colorless, odorless carbon dioxide gas that acts like a one-way mirror, trapping even more heat in the atmosphere. If coal should become our prime energy source once oil is depleted, the global heating fiasco could be frying our descendants within less than a century.

As if burning coal weren't bad enough, humans also directly pollute the air with carbon dioxide. Every time we breathe, this gas exhales from chemical reactions within our bodies, the result being that a burgeoning populace itself contributes to global heating. Plants can absorb only so much carbon dioxide. And with logging operations increasing to provide the housing and paper needs of a mushrooming population, vast forestry absorbers are steadily falling by the wayside. In point of fact, the carbon dioxide content has recently been measured to be on the rise. There's no argument about it.

After millions of years of being subjected to the whims of the environment, humans are now gaining the know-how to change that environment. But heating it is not the way to do it.

Two further notes: Some researchers argue that human-induced heating might be offset, ironically enough, by further degrading the quality of the air. In theory, microscopic debris is capable of reflecting sunlight, thus cooling the biosphere. In practice, however, the bulk of the debris naturally released as by-products of industrial production is soot or smog. Instead of reflecting solar radiation, dark and dirty soot absorbs much of it, further heating the biosphere. So it would be a mistake to grant big business a license to pollute liberally under the guise of quenching atmospheric heating.

Other critics offer more exotic arguments, claiming, for example, that biospheric heating will be countered by a natural cooling experienced as our planet oscillates back toward an ice age. This is not a valid argument, however, for the time scales are not even approximately comparable. The next ice age isn't expected for many thousands of years. Over the course of the next several centuries, or even the next thousand years, Earth's temperature will not decrease appreciably as a result of natural causes. The astronomical alignment that triggers planetwide cooling is too far in the future. Let's not rely on the coming ice age to bail us out of this thermal pollution problem.

The bottom line is that it would be a fatal delusion to think that we can generate unlimited amounts of energy to support the daily needs of a rapidly multiplying population. Being bullish on economic growth is not a solution to our problems; bullishness itself will create new problems. To avoid the unhealthy heating of our environment, the total energy use of our civilization must soon level off. This can be achieved only by checking the growth of world population.

Change or perish.

∞

Aside from problems caused by large numbers of people on planet Earth, potentially grave troubles lie ahead because of the actions of just a few individuals. Designed self-destruction is a case in point here — it could conceivably cause the obliteration of intelligent life.

Modern warfare is an especially germane example of self-destruction. Military organizations are constantly developing new ways to kill people and destroy all manner of things. In 1980, the United Nations announced that the nations of the world spend a million American dollars *per minute* on weaponry. Nuclear explosives, laser-guided weapons, neutron bombs, mobile missiles, lethal chemicals, and a growing arsenal of other destructive devices have become permanent ingredients of our civilization. These are not just popgun fare, capable of maiming individuals; they are global munitions, capable of mangling whole nations. Consider nuclear bombs.

The world supply of nuclear weaponry is currently equivalent to some twenty billion tons of TNT, a highly explosive chemical used in the production of dynamite. Numbers in the billions no longer faze readers of the earlier chapters, but a consideration of weapons density is guaranteed to shock anyone. Dividing this world arsenal by the number of people now on the planet, we find to our astonishment a sum total of five tons of TNT *per person*. This is neither five bullets, nor five sticks of dynamite, but the nuclear equivalent of five tons of explosives for every man, woman, and child on Earth. No wonder it's called overkill!

Further reflection reveals the extent of our disgrace, not just because we pay for all these armaments, but especially because we tolerate them. We are members of a society that permits the unchecked escalation of nuclear arms that can be used for one thing — to wage nuclear war. And, contrary to popular belief, the Strategic Arms Limitation Talks do not reduce this weaponry. At best, this bilateral lip service acts only to regulate the expansion of world-destructive powers.

What sort of damage does a typical one-megaton nuclear blast guarantee? A hundred times as destructive as the Hiroshima bomb, the detonation of the equivalent of a million tons of TNT would create a brilliant fireball, the center of which would attain temperatures of millions of degrees Celsius, comparable to the Sun's interior. Such rapid heating causes sudden expansion of the air around the point of explosion, which in turn gives rise to a shock wave or severe wind where pressures reach values several thousand times that of Earth's normal air. This extreme pressure would be sufficient to flatten ordinary brick houses some four kilometers (2.5 miles) away from point of impact. One such typical nuclear warhead, of which there are now tens of thousands in the world's arsenals, would be absolutely fatal to buildings, people, and almost everything within a fifty-square-kilometer area surrounding ground zero. Not only would the blast be grossly destructive, but the heat released by a one-megaton nuclear explosion can cause paper to ignite as far away as fifteen kilometers (ten miles), ensuring widespread fire storms throughout the region. The destruction of life and property would be so immense, regardless of where in a city such a weapon landed, that missile accuracy isn't even required.

This description is not offered to elicit hysteria. These are facts, bold, stark facts. Construction of nuclear bombs is based on the laws of physics, and the destructive aftermath of their use is also dictated by those same laws.

Furthermore, it's important to recognize that nuclear bombs are not just scaled-up versions of conventional armaments. The radioactive particles produced during the explosion itself, as well as those destined to fall out of the atmosphere far beyond the impact point, would cause virtually irreparable damage, rendering the land useless for hundreds, perhaps thousands, of years. Clearly, a major nuclear war would leave the face of our planet drastically changed, perhaps uninhabited. It's likely that everything we cherish as great and beautiful would be lost.

In a world of such enormous firepower, there can be no true defense. The United States and the Soviet Union both harbor terrible destructive forces, and each side knows the other side has them. The outcome is supposedly a "stable" situation where neither country would dare strike — an equilibrium called by some a balance of power, and by others, peace by fear. The catch-phrase in the language of Pentagonese is "mutually assured destruction," the Strangelovian acronym for which is MAD.

Alternative approaches, like the one of strategic targeting of only military sites recently announced during the 1980 American election year, are unlikely to deter wholesale destruction in the event of war. It just may enhance the chances for war.

In any event, the real state of affairs is not complete stability. Every so often, instabilities arise to enhance the chance for war between the superpowers. Such an instability might result from an international crisis, perhaps directly involving the United States and the Soviet Union, or perhaps initially entangling other countries and eventually escalating to the point of threatening military conflict between the superpowers. Instability could result from short-term confrontations like the Cuban missile crisis in the early 1960s and the six-day Middle East war in 1967, or from long-term hostilities like the protracted Vietnam affair and the lingering revolutions in the Persian Gulf area. Though unexpected international conflicts don't make outright war a certainty, they surely don't increase the probability for peace, either.

More predictably, instabilities in the balance of power regularly occur as major weapons systems either are introduced or become obsolete. For a certain period of time, one side has or thinks it may have a slight advantage. Upgraded weaponry may even grant to one side a first-strike capability, whereby one superpower could launch an attack so devastating that the other government wouldn't be able to respond offensively. For example, construction and deployment of "smart" cruise missiles, mobile-launched nuclear bombs, or multiple independently targeted re-entry vehicles are thought by some to give the United States a decided advantage, at least until such time as the Soviet Union can neutralize these new weapons with countermeasures of its own. Likewise, the introduction of a whole new class of Soviet intercontinental ballistic missiles having enormous throw weight, or the development of killer satellites as part of their modern armory, is often regarded as giving them a net advantage — at least until such time as our government unveils yet other new weapons sufficient to reestablish the power balance. Even the recent Russian attempt to build massive underground shelters, which would protect key elements of their civilian and industrial centers from nuclear conflagration, serves to upset the balance of power. A national Soviet civil-defense program could be viewed as a Soviet advantage or at least an instability, since the United States would no longer be able to hold the Russian populace hostage, as the Soviets presently do the American population in the absence of a significant United States civil-defense program.

There are numerous other examples of international crises and weapon developments, all of which serve to enhance the probability for war simply because the superpower scales are slightly imbalanced. These are among the main issues at the Strategic Arms Limitation Talks in Geneva and at the United Nations sessions in New York City. Yet, the arms race continues virtually unabated, and ugly international confrontations flash repeatedly across our globe.

What effect will a recurrent series of instabilities have on the future of our civilization? Well, the outlook seems to be inevitable nuclear holocaust in the Northern Hemisphere. To see this, consider the following analysis.

Suppose that, on average, there is an instability every half-dozen years. This is roughly the frequency of major rifts in the balance of power since the Second World War. Suppose furthermore that there is a ninety-five percent probability for peace during any one period of instability; that still leaves the odds at one chance in twenty, or five percent probability, for the outbreak of full-scale war. Then inquire about the degree to which the compound probability for war increases as civilization navigates through several periods of instability. In other words, how many periods of instability can a technological civilization withstand before the total probability for war exceeds the total probability for peace? The answer is about seventeen such instabilities, or a hundred years.

Self-destruction via modern warfare is, of course, always possible at any time, even when the superpowers are evenly balanced. No one really knows the exact chance for war during stable times, though we might imagine it to be very small. Computations like the one above simply suggest that the probability for war not only increases a little during any one period of instability, but also grows steadily throughout the course of time, each chink in the power balance not being entirely independent. If, according to the above estimate, global instabilities raise the chance for war to five percent at any given time, then the compound probability for nuclear holocaust becomes higher than that for peace — namely, fifty-one percent — after only ten decades. Should this sterilized examination of war and peace in any way approximate the truth, then we're roughly a third of the way to Armageddon.

Should the average probability for war during periods of instability be greater, then this type of analysis suggests that nuclear war could be imminent. Conversely, a lower probability for war during individual instabilities suggests that the superpowers might be able to avoid war for a longer time. Only if war's probability in these circumstances is far less than one percent can we hope to postpone nuclear catastrophe for more than a few centuries, a time scale still small in the cosmic scheme of things. No matter how minute the chance, though, the compound probability for war will sooner or later exceed that for peace, making full-fledged nuclear war better than a fifty-fifty shot.

The crux of any analysis like this one revolves about the average probability for war during any one instability. Of course, no one knows this value for sure. There are too many factors involved, including the nature of humanity, which doubtless influences in some complex manner the response of governments either to trigger or to avoid nuclear self-destruction. Numerous sociopolitical factors would seem to play an integral role, but none of them can be quantified, and at any rate, in a rapidly advancing technological society, these factors may become nearly irrelevant. The nature of humanity, moreover, might not play a role even in times of crisis. If not, then the argument is clear solely on the basis of probability theory, and it is this:

Though the probability for war may be small during any single period of instability, a civilization can withstand only so many instabilities before the compound probability for war begins to exceed that for peace.

Of course, there are always opponents of this type of analysis. Nor are they necessarily those permanently equipped with rose-colored glasses. They argue, for example, that our leaders wouldn't actually retaliate, even knowing that some other government's nuclear arsenal was due to arrive within the twenty or so minutes that intercontinental ballistic missiles need to travel from one point on the globe to any other. But how can we trust any leader of a government to sacrifice its people for the good of civilization? Retaliation is becoming so mechanized that the element of humanity is minimized, perhaps lost. If certain elements of our "Defense" Department have their way, the American response will soon be triggered only by radar-computer systems, not by the President. This elected civilian can override the computer by vetoing the command, but in the event that he or she hesitates, our missiles will be up and away — automatically. We live at a time when command and control decisions are being transferred to machines. And, for better or worse, words like "humaneness," "civilized," and "survivability" just do not compute.

Others maintain that nuclear weapons will never be used. Retaliation, they claim, is not an issue because no one will be foolish enough to unleash nuclear weapons in the first place. But how can we afford to believe this viewpoint? That's all it is — a belief. Warring nations have seemingly never failed to utilize the most potent weapon available to them. Since early history, the buildup of weapons and the prospect of war have been closely allied; the development of arms has virtually always precipitated their use in warfare. With few exceptions, each new and deadlier weapon, from crossbow to guns to dynamite to tanks to atomic bombs, has eventually been deployed on the field of battle. Historically, once humans have fashioned a new weapon, they apparently commit themselves to its use. Those who say that nuclear weaponry is different deserve a cold response: How can we be so sure that the goodness, the rationality of humanity, will surface in the nick of time?

Still others argue that, despite a full-scale nuclear holocaust, all inhabitants of planet Earth will not necessarily perish. This is tan-

tamount to saying that nuclear war is winnable. But the very concept of assured destruction is designed to make this impossible.
With so much overkill now stored in our nuclear depots, it's
equally likely that any humans surviving the targeted blasts would
succumb to the postwar aftermath of radioactive fallout, economic
chaos, ozone depletion, and climatic cooling. The cumulative effects
of all-out nuclear war would be so catastrophic that they render
any notion of "victory" meaningless. Such arguments are totally
specious, offered by irresponsible people — the type, unfortunately, who have thus far generally overseen the design and development of nuclear weapons policy. Let's once again hope that
pronouncements like these are not true reflections of the real
nature of humanity.

The currently misleading concept of mutually assured destruction must be reframed in more realistic terms to reflect the full
magnitude of the cataclysm that a nuclear war would represent.
Then, we can plan, not to limit nuclear weapons, but to ban them
altogether. So, let's change the philosophy: Instead of arguing for
mutually assured destruction, we should strive to reach a loftier
goal of mutually assured survival.

The above predicaments are surely a good deal more complicated
than here sketched, largely because several of them are interrelated. For example, should current sociopolitical attitudes remain
unaltered, the chances that someone will unleash nuclear bombs
— that's the self-destruction problem — will surely increase as the
number of inhabitants grows — that's the population problem.

Current conflicts are destined to become further inflamed as
people, perhaps whole nations, become desperate for food and
energy. Wars waged solely to redistribute wealth may be the only
way that poor nations, which feel they have nothing to lose, can
hope to remedy their deteriorating status. The prospect that developing nations might catalytically induce nuclear war between
the superpowers grows steadily. Even the specter of ski-masked
terrorists conducting nuclear blackmail with clandestine plutonium devices looms large on the horizon. Doubters should keep

in mind that significant quantities of plutonium and enriched uranium, produced in American nuclear plants, are currently missing. Furthermore, at least one large American city has already seriously considered capitulating to a multimillion-dollar demand on the threat that the city would be leveled by a hydrogen bomb, a hoax which, at the time, neither the Atomic Energy Commission nor the FBI could discredit.

Clearly, the continued growth of world population and the incessant threat of nuclear holocaust are the foremost problems confronting our civilization today.

Change or be doomed.

∞

Let us assume that we can solve all the above problems, recognizing that the severity of each one can be lessened through shared information, intelligence, foresight, and change.

Even so, there will be other global annoyances. Once any civilization moves beyond technological adolescence, it will regularly and naturally encounter all sorts of hazards while evolving toward the future. Though less certain than those noted earlier, such crises are bound to have some debilitating effect on the well-being of Earth civilization. Consider a couple of conceivable difficulties — a decrease in the quality of genes, and an increase in the quality of computers — each of which could be upon our descendants sooner than most of us are willing to admit.

Genetic degeneration is a good example of a futuristic problem that civilizations must circumvent, suppress, or otherwise resolve in order to survive. This is a basic dilemma, like the population explosion, stemming from the discovery of bacteria. Like the population problem that mushrooms rapidly once it reaches the explosive portion of the exponential curve, genetic degeneration can subtly infiltrate communities of technologically intelligent creatures before bursting into the open. It will not be easy to recognize before our descendants find themselves pinned down and in desperate straits for a fix to help solve it. Futuristic or not, the roots of this problem seem to be taking hold in our society right now.

Recall for a moment the central dogma of neo-Darwinism: Accidental alterations in the DNA structure permit organisms to respond to a gradually changing environment. For some, the adaptation enables them to survive famously, often carving out whole new niches; for others, the adaptation is not so successful, causing them to become dead-ended or even extinct. In short, genetic mutations followed by natural selection prompt survival of the fittest. This is the rule of order for all living things. Given enough time, though, even nature's rules change.

Once a species evolves technological intelligence — at least the competence to invent medicine — the infant mortality rate decreases, the average life span increases, and not only do the fittest survive but almost everyone else does too. Although medicine cures contagious disease at present, it would also seem to promote hereditary disease in the long run. Expressed in another way, medicine currently improves health for us, but it could lead to a genetically polluted society for our kids' kids' kids' kids' kids.

The gene pool is shifting. Mentally and physically deficient humans, previously unable to survive to the point of reproducing, now yield fertile offspring. Surely they contribute favorable as well as unfavorable traits, but the upshot is that the distribution of active genes within the human species is now rapidly changing. No mutations are involved in this change; medicine alone is the culprit.

Can the normal course of biological evolution keep pace with this rather sudden change in the availability of genes induced by the invention of medicine? The answer might be in the affirmative if biological evolution were indeed occurring as it has for the past several billion years. However, the classical biological evolution of men and women is coming to a halt. Why? Because an even more fundamental change is now under way.

After more than ten billion years of natural evolution, the dominant species on planet Earth is beginning to tinker with evolution itself. Whereas previously the gene (DNA) and the environment (stellar, planetary, geological, sociological) governed evolution, now we are suddenly gaining control of both the environment and the gene. We now tamper with matter, diminishing our planet's resources, often polluting it all the while. And we stand on the threshold of tampering with life, potentially altering the genetic makeup of human beings.

Some specialists, seemingly impervious to long-term generalities, argue that genetic degeneration will not materialize into a global crisis threatening the survivability of civilization. They base their assessment on the presumption that our descendants will not hesitate to modify, through microsurgery, the DNA structure of certain types of individuals. In other words, the current slippage in the gene pool toward the "bad" side can, in principle, be checked by unchancy, premeditated alterations of the genetic molecules themselves.

For some time now, science fiction writers have had a name for this kind of operation: breeding. The whole affair is only now beginning to convert from fiction to fact. Once the transition is under way, the full thrust of the dilemma along with its vast ethical implications will burst forth. Who controls the breeding? Which species, races, individuals should be bred? What's to prevent the cloning of whole armies of professional brutes, however remote a possibility? How might genetic engineering affect the decisions necessary to avoid nuclear and other types of self-destruction? Even if we could extract certain traits — aggression, perhaps — from the human psyche, *should* we do so? Without aggression and its competitive edge, what might happen to our exploratory drive, our curiosity?

The increasing sophistication of "smart" machinery — exemplified by the computer — is another potential hazard with which our descendants must eventually come to grips. Might artificial intelligence be the next stage in the ascent of cosmic evolution? Is it only a matter of time before silicon-based computers surpass or even replace carbon-based humans? Are our days of intellectual dominance numbered?

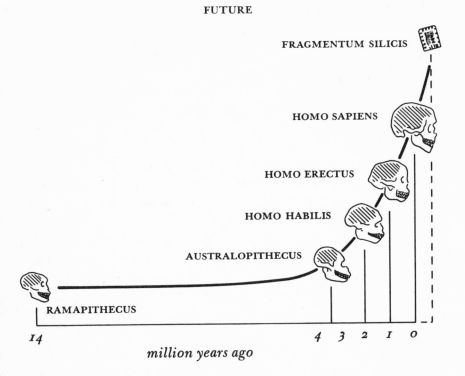

FUTURE

FRAGMENTUM SILICIS

HOMO SAPIENS

HOMO ERECTUS

HOMO HABILIS

AUSTRALOPITHECUS

RAMAPITHECUS

14

4 3 2 I 0

million years ago

Scientists are now attempting to program computers to think, learn, and even create. The extent to which artificial intelligence might be advanced is not yet known. Though some researchers maintain that today's computers are nothing more than devices to serve our human wishes, everyone agrees that such machines can now perform sophisticated mathematical calculations much more quickly than humans. Speed is truly their forte. In addition, computer memories have grown larger, enabling them to store vast quantities of raw information.

Programs written by humans, fed into today's most advanced computers, and further refined by the machine's memory, have now enabled those computers to beat any human at checkers, and almost any human at chess. Having made a blunder, a computer stores that mistake in its memory, never to be repeated; nonliving computers, perhaps more effectively than living beings, can learn from experience. When the allowed time limit on moves is made short enough, computers can easily trounce even a chess grand master. Furthermore, robots capable of making limited choices on their own without constant guidance from Earth are now being built to explore the surfaces of other planets of our Solar System.

Debates now rage about whether machines can ever be built to think or understand completely on their own. If learning is possible, is creativity next? Will computers someday not only answer key questions, but also ask them? Some researchers suggest that such devices can never be constructed, that there are fundamental differences between humans and machines that no amount of technology can overcome. These critics claim that it's incorrect even to suggest that computers work like human brains. The human qualities of will, intuition, emotion, and consciousness, not to mention the oft-claimed significance of a "mind" separable from the brain, all tend to suggest that today's computers are, by comparison, nothing more than incredibly fast morons. Others disagree, regarding thinking computers as inevitable, given the rate of technological innovation. They believe that humans are essentially machines who will eventually build better machines to replace themselves.

In any event, should intelligent machines ever really make us uncomfortable, it would seem that we could always "pull the plug." Still, it will be important to monitor the effect on human values, human ethics, and self-respect in a world where people *think* computing machines might someday dominate us.

∞

Our future, the future of the planet, is cloudy, shrouded in numerous dilemmas, many of which threaten the continued existence of Earth civilization and perhaps Earth life itself. Can our precarious condition persist indefinitely? What is the longevity of a technologically competent civilization?

We stress that the problems facing civilization today are not similar, not even in principle, to those of previous generations.

The recent exponential expansion of technological achievements and the inability of society to cope with them have led to *global* problems basically different from those confronting earlier peoples on Earth. In no other time in history have humans commanded the means to affect the status of large segments of life on our planet. In one way or another, it's not inconceivable that the entire planet could be rendered lifeless in the near future.

Examined from afar, the whole predicament before us seems paradoxical. Evolution of matter is responsible for our Galaxy, Sun, planet, and life. And evolution of life is responsible for our intelligence, culture, and technology. But now this very same phenomenon — change — seems to be threatening us. Change has caused many of our present problems, yet to circumvent these problems we need more change. The cause of it all is that humanity is in the process of sliding into the driver's seat. Humans now effect change more rapidly than nature could ever do.

In the past century, we have increased our speed of communication by more than a million times, our speed of travel a hundred times, and our energy consumption a thousand times. Within the past couple of decades alone, we've enhanced by several million times the speed of data processing and the efficiency of weapons development.

Some people, unable to cope with rapid change, argue that technology is the principal cause of many of today's troubles. Exponential population growth, environmental pollution, depletion of natural resources, shortages of food and energy, threats of nuclear incineration, the potential for genetic degeneration, and a host of other ills are now, or soon will be, threatening the viability of Earth civilization and quite possibly all of Earth life itself. Some of these problems are in fact by-products of technology. Yet the saddest problem of all is that our social and political organizations seem unprepared to provide the innovative, foresighted responses needed for our continued existence.

For better or worse, we can be sure that our civilization is fast approaching finite natural limits. There is now a fundamental change in the way things change — at least on Earth. This does not necessarily mean an end to civilization. But we are doubtless approaching a finale for the rapid change brought upon ourselves. We cannot communicate faster than light velocity, a speed already achieved by radios and televisions. We cannot travel around the Earth faster than orbital velocity, an ability already attained by space probes and astronauts. We cannot solve the population glut by copping out into space. We cannot consume fuel to the point of thermally polluting the air and melting the polar ice caps. And when it comes to nuclear weapons, we cannot be deader than dead.

Our civilization is now in transition, the likes and scope of which no Earth society has ever before encountered. What are we to do? There are only a few options.

First, our fast-paced society can fail to solve any one of the global problems soon to face us, in which case, for all practical purposes, civilization goes down the tube. This is the view of the doomsayers. Out of the remnants of such a dead civilization, another one could perhaps arise anew. Perhaps, but perhaps not. In any event, this "solution" concedes that advanced civilizations are unlikely to survive very long after crossing the threshold of technology.

Second, civilization may successfully resolve each of Earth's problems as each in turn becomes critical. Resembling a person navigating a minefield, we can take one step at a time, using technology to lighten our burdens, to help us progress and thus survive. This is the view of the technological optimists. This solution seems reasonable at face value, but deeper thought suggests a fundamental dilemma.

To avoid any of our impending worldwide troubles, civilization must be willing to sacrifice. To avert, suppress, or otherwise alleviate any one problem requires our civilization to exercise some restraint. Should our descendants elect to choose this second route into the future, then our species is destined to become less free to do what it wants, more constrained to do exactly whatever is necessary to guarantee survival. This second route aims us directly toward a state hallmarked by regimentation — a state hard to define but one where personal freedom, dignity, curiosity, and many of the other cherished qualities of human nature are diminished, perhaps even eliminated. Complete stability would seem to imply a stagnated, unprogressive state where human rights are not only absent but perhaps not even understood.

We might then wonder: If curiosity dies, does intelligence die also?

Let's face it, continued development of our civilization — and thus any further attempt to extend technological longevity — will require a delicate balance between opposing hazards, each of them unacceptable. On the one hand, we face the danger of catastrophe if even one global problem goes unsolved. On the other hand, we face the temptation of becoming a stagnant society as a result of the increased stabilization and regimentation needed to ensure survival.

After billions of years of evolution, life on Earth has finally arrived at the greatest possible dilemma for any neophyte technological civilization. Will our generation — humankind at the turning point — make some contribution to the long-term survival of the species? Are we willing to face the issue of survivability of humankind in a world threatened both by physical extermination and by various forms of dehumanization? Is there no narrow path between the danger of catastrophe and the temptation of stagnation?

Some scientists now suggest that there is a recipe for avoiding stagnation, while still surviving. The vehicle necessary to guide us along the intermediate course may well be a twofold program of simultaneously colonizing the nearby planets and searching for galactic civilizations. Such a project may sound far out — and, literally, it is — but sober reflection suggests that prescient outreach of this sort makes more sense than bland, inward-looking pronouncements offered under the guise of raising social consciousness or reviving religious fervor. Turning inward, while trying to halt change, cannot indefinitely preserve our species on this small planet; looking outward, while accepting change, should now be embraced as the means to our coming of age as an intelligence in the Universe.

Humans can and should begin to terraform and colonize other planets and their moons. This is not tantamount to abandoning ship. Nor is this a wild-eyed scheme to look to the stars for solutions to our Earthly problems. (Indeed, our present double-trouble potential for population glut and nuclear war must be checked by changing the social climate here on Earth, and quickly at that.) Instead, this is a rational program that mostly transcends our present problems, while at the same time identifying a common objective to divert both sides' military-industrial complexes from misusing vast amounts of human resources. Of greatest importance, a program of partial movement away from our planet would enable us to disperse Earth's civilization throughout larger pieces of interplanetary real estate.

Recall that we earlier criticized space colonization. *Planetary* colonization is a demonstrably different proposition. Terraforming, the process of transforming planets or moons into Earth-like objects, may at first sound ridiculous, yet it's probably cheaper, safer, and more realistic than constructing monstrous interplanetary bottles for people to inhabit the insides of. Planets or their moons provide solid foundations for habitats. And some of them, Mars and Venus, for example, have enough mass to retain their own atmospheres, thereby enabling people to reside naturally on the outside.

Admittedly, substantial environmental changes would be required to transform the mostly carbon dioxide atmospheres of these planets into breathable oxygen. But there are known natural processes that could help. For example, the seeding of copious amounts of blue-green algae could convert their alien atmospheres, via photosynthesis, into oxygen-rich environments.

The prime motive for planetary colonization is the dispersal of the human species. Any of the above-discussed global problems is a real danger to societies confined to a single planet. But once a society scatters over astronomical distances, then it becomes a good deal less vulnerable to local catastrophe. Invulnerability — that's the key to the survival of a technological civilization. Individual colonies might fail to survive, especially if they encounter many of the ailments now or soon to be confronting twentieth- and twenty-first-century Earthlings. But the chances are lessened that all such colonies would so perish. Even if only one planetary colony were to survive the onslaughts of terrestrial and extraterrestrial catastrophes, that would be enough to preserve our civilization, our species.

Now is the time to inaugurate a dedicated, preferably international effort to lay the groundwork for planetary colonization. The more quickly we begin to exploit the natural resources of our Solar System, thereby converting some of its matter into biological living space, and especially scattering our kind throughout a large volume, the better the chances that Earth's great experiment — intelligent life — will not end in dismal failure.

The second program that humans should now undertake is a search for galactic civilizations. What are the chances of success? Well, recall that we've come to recognize that we inhabit no special place in the Universe. All experimental tests made since Renaissance times point toward the idea that we are residents of an undistinguished hunk of rock, circulating about an average star, someplace in the suburbs of the Milky Way Galaxy. If we are examples of anything in the cosmos, it is probably of magnificent mediocrity.

With one exception, there's nothing unique about planet Earth. The exception is that Earth is the only place in the Universe where we know that life definitely exists. We might prefer to think that cosmic evolution operates everywhere in the Universe but, to be truthful, we know of no other location in the entire Universe where life has arisen. This doesn't imply that life is absent beyond our planet. It means that if extraterrestrial life does exist, we haven't yet become sophisticated enough to know it.

The case favoring the prospects for extraterrestrial life can be summarized by noting what are sometimes referred to as the assumptions of mediocrity. In particular, one can make the following general argument: Since life on Earth depends on just a few basic molecules (in fact, molecules composed of atoms common to all stars), then if the laws of cosmic evolution as we know them apply to every nook and cranny within the Universe, life may well originate at numerous places throughout our Galaxy and galaxies beyond. In other words, given the vastness of the Universe, and the astronomically long time scales involved, it seems unreasonable that life could have arisen only on planet Earth. Even if planets having congenial environments for life are as rare as one in a billion star systems, there will still be at least several hundred such candidates in our Galaxy alone.

The opposing view maintains that intelligent life on Earth is the product of inconceivably fortunate accidents — astrophysical and biochemical mistakes that are unlikely to occur anywhere else in the Galaxy or the Universe. Cosmic evolution is not challenged. In fact, it's embraced as correct in this alternative view. Life it still considered to be a natural consequence of the evolution of matter; it's just not an inevitable consequence. Accordingly, the steps leading to life — especially intelligent life — are in this view considered to be so unlikely as to make it unreasonable to conclude that there's anyone else out there.

Researchers subscribing to this latter view argue that a search for extraterrestrial beings is unreasonable and unwarranted. They claim that the assumptions underlying the prospects of extraterrestrial intelligence contain too many uncertainties. The search strategies themselves contain additional uncertainties. They conclude that any expenditure of time, effort, or money for a search is unjustified by the meager evidence at hand.

Proponents of a search for extraterrestrials admit there is only a slight chance of making contact in the near future. But they contend that now is the time to test the hypothesis that other technological civilizations inhabit our Galaxy. Failure to do so is tantamount to committing the cardinal error of pre-Renaissance workers — thinking without experimentally testing.

Citizens of Earth now stand on the threshold of membership in

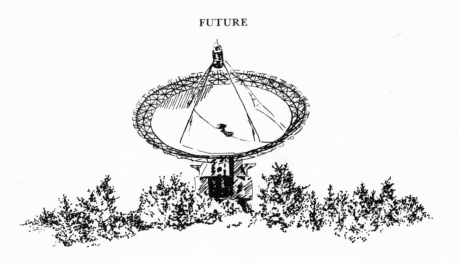

the community of galactic civilizations — provided we really want to join it, and provided there really are galactic civilizations in the depths of space. With the aid of modern experimental gear such as radio telescopes, space probes, and digital computers, our civilization is now capable of taking the next great evolutionary leap forward — making contact with extraterrestrial intelligent life.

Assuming that such a galactic presence exists, we might profit enormously by communicating with it. We have, after all, only decades ago acquired a technological competence, only recently gained the ability to engage in an interstellar dialogue. Accordingly, we are quite likely to be the dumbest intelligent civilization in the Galaxy.

At the least, discovery of advanced galactic life will assure us that it is possible for technological civilizations to avoid doomsday — to survive.

At the most, by establishing an interstellar dialogue, our civilization may well be able to strive toward a higher level of consciousness heretofore unimagined.

This is not to suggest that contact itself will provide us with instantaneous intelligence, though it might. Nor do we suggest that extraterrestrials will be able to provide solutions to our global predicaments, though they might. The suggestion now being made by numerous researchers is that the very program of searching

293

will stretch our imagination, widen our horizons, enhance our curiosity. The search itself becomes humankind's instrument of survival.

The hallmark of humanity is exploration — an innate desire, an insatiable thirst, to know who we are, where we came from, and where we're headed. Exploring means observing: watching, probing, questioning. To halt exploration at this juncture would be to act against the very attribute that makes us human. Failure to inhabit nearby matter, and at least to reconnoiter distant matter, might prematurely terminate humankind's exploratory drive. Indeed, civilization's technological longevity might prove to be short simply because we didn't undertake the colonization of planets and the search for extraterrestrials.

One thing is worth remembering: The space surrounding all of us may be, right now, inundated with radio signals broadcast by extraterrestrial civilizations. If we only knew the proper direction and frequency, we might be able to make the most startling advance since the birth of language. The result would not only be a whole new vista in the study of the evolution of matter and life, but it might also be the key to our civilization's survival.

An international effort at planetary colonization and extraterrestrial searching would provide additional stimulation and competition needed for survival. It would boost our curiosity to enormous heights. And it would enable us to postpone, perhaps indefinitely, the regimented, totalitarian society for which we now seem destined.

The critical concern for us in the years ahead is this: When a technological civilization tries repeatedly to solve the numerous planetwide crises that inevitably confront any evolving society and, in doing so, plunges straight toward mental and physical stagnation — the crisis that ends all crises — is there enough time to disperse into its planetary system and to establish an interstellar dialogue? Human evolution of this magnitude has in the past usually required tens of thousands of years, though there are some precedents for significant change over much shorter time periods:

the transition from hunter-gatherers to agriculturists, and that from feudal to industrial society. Clearly, enormous financial enthusiasm and social commitment will be required to sustain such a duplex project of terraforming and searching for decades, perhaps centuries.

We may have evolved in the past from Universal matter, but our future has been largely delivered into our own hands. Are we smart enough to adjust to this alteration in the evolutionary scenario? Are we wise enough to ensure the survival of humankind? Our destiny will truly be a measure of our current intelligence.

A WHOLE NEW ERA

billions of years ago

THE SCENARIO OF COSMIC EVOLUTION is a human invention. Despite seven major themes, it was not handed to us on a stone tablet.

Nor is the idea that we are children of the Universe a new one. That notion may be as old as the earliest *Homo* to contemplate existence.

But while approaching the end of the twentieth century, we can begin to identify conceptually some of the subtle astrophysical and biochemical processes that enable us to recognize the cosmos as the origin and source of our being. The approach is very much inter-disciplinary, interweaving knowledge from virtually every subject universities offer. And — of great import — many parts of the grand scenario sketched here have recently been substantiated by experimental science.

Cosmic evolution is a broad working hypothesis that strives to integrate the big and small, the near and far, the past and future, into a unified whole. Though several key details remain outstand-ing, the overall framework of existence, including its steady flow from radiation to matter to life, seems to be comprehensible.

Let us once more examine the broadest view of the biggest pic-ture. In the earliest epoch of the Universe, radiation dominated matter. During this Radiation Era, nothing could have been ob-served except intense light. The vast amounts of radiant energy produced at the time a spectacularly bright fireball inside which no atoms or molecules could have formed.

As the Universe expanded, it cooled and thinned. Having evolved from radiation, matter gradually began coagulating into atoms and eventually into clusters of atoms. Indeed, an event of incomparable significance occurred when matter first began to coalesce shortly after the Universe originated in a blinding maelstrom. The emergence of matter as the dominant constituent is the first great transformation in the history of the Universe. This change was fundamental, an absolutely integral part of the big picture.

From the start of the Matter Era, matter then dominated radiation. And it has dominated radiation ever since, successively forming galaxies, stars, planets, and life.

Of all the known clumps of matter in the Universe, life forms are surely the most fascinating, especially those enjoying membership in advanced technological civilizations. Technologically intelligent life forms differ fundamentally from lower forms of life and from other pieces of matter scattered throughout the Universe, not only because they can tinker with matter but also because they can change the course of evolution.

Given enough time, even evolution evolves.

Be assured, elemental evolution continues unabated in the hearts of stars everywhere. Chemical evolution still occurs in remote places such as interstellar clouds and perhaps other planets. Biological evolution persists for most species on Earth and possibly other planets as well. And cultural evolution endures in many corners of our world and conceivably in other worlds beyond. But for technologically intelligent life, classic evolution seems to be undergoing fundamental change.

The emergence of technologically intelligent life, on Earth as well as perhaps other planets, heralds a whole new era — a Life Era. Why? Because technology enables life to begin to control matter, rivaling that previous transformation when matter began uncoupling from radiation more than ten billion years ago. Mat-

ter is now losing its total dominance, if only at those isolated locations where technologically intelligent life resides.

The transition from a Matter Era to a Life Era will not be instantaneous. Just as it took lots of time for matter to conquer radiation in the early Universe, long durations will surely be needed for life to best matter. Life might not, in fact, ever fully dominate matter, either because civilizations might never gain control of material resources on a truly galactic scale, or because the longevity of technological civilizations everywhere might be short.

Though a mature Life Era may never come to pass, one thing seems certain: We humans on planet Earth, along with any other technological life forms throughout the Universe, stand on the verge of experiencing the slow uncoupling of life from matter. This is a transition of astronomical significance, the dawn of a whole new reign of cosmic development.

The emergence of technologically intelligent life as the dominant constituent is the second great transformation in the history of the Universe. It is the quintessential event in the development of matter, the threshold beyond which life forms can truly begin to fathom their role in the cosmos. Significantly, then, we have an obligation, a responsibility to survive. The great experiment that intelligent life represents must not end in failure.

* * *

The cycle is nearly complete. Life now contemplates life. It weighs matter. It ponders our origin and our destiny. It explores the planetary system we call home. It searches for extraterrestrial life. It quests for new understanding.

In the process, life discovers a meaning, a relevance to cosmic evolution, an underlying motive for Universal change. Life comes to recognize that countless billions of stars were born and have died to create the matter now composing our world. We ourselves are made of matter forged in the hearts of stars, annealed in the crucible of billions of years of evolution — a kind of cosmic reincarnation. We've furthermore become smart enough to reflect back upon the material contents that gave life to us. And what we find, quite literally, is that we are more than products of the Universe, more than life *in* the cosmos. We are agents *of* the Universe — agents of the Universe commissioned by it to probe itself.

Depressing? Frightening? Absolutely not. Cosmic evolution is wonderfully warm and enlightening, enabling us to embrace our cosmic heritage, to make fuller use of our potentialities, and to utilize means whereby sentient beings, here and elsewhere, can now unlock the secrets of nature for the betterment of ourselves and our Universe.

Provided civilizations continue to seek new knowledge, provided they are wise enough to survive, provided above all they remain intellectually curious, then it's not inconceivable that life could someday evolve sufficiently to overwhelm matter, just as matter overthrew radiation in the early Universe. Indeed, the destiny of the Universe may well be determined not only by matter but also by the life that arises from it. Together with our galactic neighbors, should there be any, we may eventually gain control of the resources of much of the Universe, redesigning it to suit our purposes and, in effect, ensuring for our civilization a sense of immortality.

"We are brothers of the boulders, cousins of the clouds."

— *Harlow Shapley*, an American
astronomer of the twentieth century

Some Useful Sources and
Suggestions for Further Reading

An early amalgam of cosmic evolutionary ideas, Harlow Shapley's "Life, Hope, and Cosmic Evolution," *Zygon*, volume 1 (1965), page 275, may be contrasted with a contemporary Universe view, Eric J. Chaisson's "Cosmic Evolution: A Synthesis of Matter and Life," *Zygon*, volume 14 (1979), page 23.

Prologue

Lincoln Barnett, *The Universe and Dr. Einstein*, New American Library, 1957.
P. Barker and C. G. Shugart, eds., *After Einstein*, Memphis State University Press, 1981.

Particles

Edward R. Harrison, "The Early Universe," *Physics Today*, vol. 21 (1968), p. 31.
Steven Weinberg, *The First Three Minutes*, Basic Books, 1977.

Galaxies

Harlow Shapley, *The View from a Distant Star*, Basic Books, 1963.
Harlow Shapley, *Galaxies*, 3rd ed., revised by Paul Hodge, Harvard University Press, 1972.
Simon Mitton, *Exploring the Galaxies*, Scribner's, 1976.

Stars

I. M. Leavitt, *Beyond the Known Universe*, Viking Press, 1974.
Robert Jastrow, *Red Giants and White Dwarfs*, 2nd ed., W. W. Norton, 1979.
Harry L. Shipman, *Black Holes, Quasars, and the Universe*, Houghton Mifflin, 1980.

Planets

I. S. Shklovskii and Carl Sagan, *Intelligent Life in the Universe*, Holden-Day, 1966.
William K. Hartmann, *Moons and Planets*, Wadsworth, 1972.
Fred L. Whipple, *Earth, Moon, and Planets*, Harvard University Press, 1980.

Life

Sidney W. Fox and Klaus Dose, *Molecular Evolution and the Origin of Life*, Marcel Dekker, 1977.

Robert Jastrow, *Until the Sun Dies*, W. W. Norton, 1977.

George Field, Cyril Ponnamperuma, and Gerrit Verschuur, *Cosmic Evolution*, Houghton Mifflin, 1978.

Culture

H. Chandler Elliott, *The Shape of Intelligence*, Scribner's, 1969.

Carl Sagan, *Dragons of Eden*, Random House, 1977.

Edward O. Wilson, *On Human Nature*, Harvard University Press, 1978.

Future

Sebastian von Hoerner, "Population Explosion and Interstellar Expansion," in *Einheit und Vielheit*, Vandenhoeck & Ruprecht in Göttingen, 1972.

Robert L. Heilbroner, *The Human Prospect*, W. W. Norton, 1974.

John Billingham, ed., *Life in the Universe*, MIT Press, 1981.

Epilogue

Hoimar von Ditfurth, *Kinder des Weltalls*, Hoffman und Campe Verlag, 1970.

Preston Cloud, *Cosmos, Earth, and Man*, Yale University Press, 1978.

Eric Chaisson, "The Beginnings of Tomorrow," *Think*, Jan.-Feb., 1980.

Young readers may want to try:

Victor Weisskopf, *Knowledge and Wonder: The Natural World As Man Knows It*, rev. ed., MIT Press, 1979 (high-school level).

M. Kimbrough Marshall, *The Story of Life*, Holt, Rinehart and Winston, 1980 (grammar-school level).

Eric J. Chaisson is an associate professor at Harvard University, and a staff member of the Harvard-Smithsonian Center for Astrophysics in Cambridge, Massachusetts. His research centers on the radio exploration of the Milky Way, especially the formation of stars in interstellar space, and the search for extraterrestrial intelligence in the Galaxy. During the 1970s, he taught a course on cosmic evolution to thousands of Harvard-Radcliffe undergraduates.

Lola Chaisson, née Eachus, is a scientific illustrator and former astronomer.

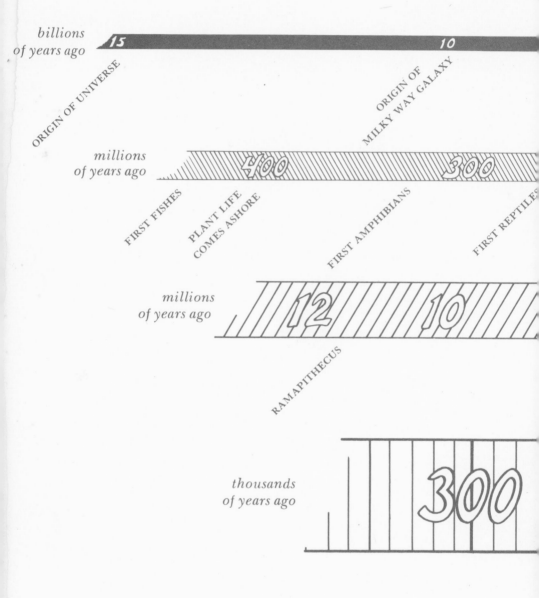

billions
of years ago   **15**   **10**

ORIGIN OF UNIVERSE

ORIGIN OF
MILKY WAY GALAXY

millions
of years ago   400   300

FIRST FISHES

PLANT LIFE
COMES ASHORE

FIRST AMPHIBIANS

FIRST REPTILES

millions
of years ago   12   10

RAMAPITHECUS

thousands
of years ago   300